THE WARHEAD

ALSO BY JEFFREY E. STERN

The Mercenary: A Story of Brotherhood and Terror in the Afghanistan War

The Last Thousand: One School's Promise in a Nation at War

COAUTHORED BOOKS

The 15:17 to Paris: The True Story of a Terrorist, a Train, and Three American Heroes

Back in the Game: One Gunman, Countless Heroes, and the Fight for My Life

THE WARHEAD

THE QUEST TO BUILD THE PERFECT WEAPON IN THE AGE OF MODERN WARFARE

JEFFREY E. STERN

DUTTON

DUTTON

An imprint of Penguin Random House LLC
1745 Broadway, New York, NY 10019
penguinrandomhouse.com

Book design by Daniel Brount

LIBRARY OF CONGRESS CATALOGING-IN-PUBLICATION DATA

Names: Stern, Jeffrey E., author.
Title: The warhead : the quest to build the perfect weapon in the age of modern warfare / Jeffrey E. Stern.
Description: New York, NY : Dutton, 2026. |
Includes bibliographical references and index.
Identifiers: LCCN 2025040787 (print) | LCCN 2025040788 (ebook) |
ISBN 9781524746421 (hardcover) | ISBN 9781524746445 (ebook)
Subjects: LCSH: Texas Instruments Incorporated—History |
Precision guided munitions—History |
Precision guided munitions—United States—History |
Vietnam War, 1961–1975—Aerial operations, American |
Persian Gulf War, 1991—Aerial operations, American |
Iraq War, 2003–2011—Aerial operations, American
Classification: LCC UF510 .S74 2026 (print) | LCC UF510 (ebook)
LC record available at https://lccn.loc.gov/2025040787
LC ebook record available at https://lccn.loc.gov/2025040788

Printed in the United States of America
1st Printing

To Dad

For roots and wings, and for everything else

Contents

Author's Note
xi

PROLOGUE: JOE
(Project Anvil)
1

BOOK I: HILTON
(Operation Rolling Thunder)
15

BOOK II: WELDON
(Operation Linebacker)
49

BOOK III: WIEGAND
(Operation El Dorado Canyon)
87

BOOK IV: KATHY
(Operation Desert Storm)
141

BOOK V: RADICA
(Operation Allied Force)
185

BOOK VI: LUIS
(Shock and Awe)
225

BOOK VII: SALMAN
(Operation Unified Protector)
297

AFTERWORD: ELKAN
347

Acknowledgments
359

A Note on Sourcing
361

Notes
363

Index
393

Author's Note

The inspiration for this book came during a single astonishing encounter I had several years ago.

It was the fall of 2018 and I was very far from home, covering a conflict for *The New York Times Magazine* in a place where everything felt unfamiliar. I'd spent more than a decade traveling to and sometimes living in war zones, but there—northern Yemen, in the midst of civil war—I was beginning to feel like I might be a stranger in ways I didn't even comprehend.

To get there I'd had to travel in disguise. There'd been a hurried parking lot rendezvous down in the southern port city of Aden, where three young men dressed me up in local clothes, wrapped a scarf around my head, and shoved a fistful of psychoactive *khat* leaves into my cheek. They bundled me into the back of a 4x4 and told me not to speak as we tried to survive checkpoints with ever more adolescent and twitchy-looking gunmen. We rode half the width of the country, crossed into rebel-held territory, and then pulled off for cover under a stand of thirsty-looking trees, and I was passed to another team of wartime entrepreneurs, who packed me into a different 4x4, and spirited me the rest of the way north to spend the night in a city under siege.

The next day we drove farther, trying to find a village in a remote region where the incident I'd come to write about had taken place.

When we finally arrived, an interpreter and I walked out across an endless valley of black rock to what had once been a water-drilling site, and I began interviewing villagers.

But I was distracted. Off in the distance there was this boy who kept leaning down and combing his fingers over stones, as if collecting seashells. Every once in a while, he'd pick one up and inspect it. At some point he walked up to me, touched my arm, and, without saying anything, handed me three he'd collected.

They were heavier than I expected. Notably, *exotically* heavy. I'd been told the terrain here was volcanic; I was thinking perhaps there was some interesting chemical explanation for why even the rocks felt foreign to me, but then the boy said something to the interpreter. The interpreter paused. Something moved across his face. He turned to me and explained that it was actually shrapnel. I was holding part of a bomb.

There must've been something in my own expression, and one of the village men standing nearby must have seen it, because a little later, we all gathered in someone's living room, this one villager followed me there, and when we were all sitting on the carpet, he asked me to touch his face. He leaned forward, chin-first, and cocked his head as if about to wink. He said there were also pieces of metal still lodged in a lip of skin beneath his eye. I paused and he nodded as if to say, *Go on.*

So I reached out my hand. And in this instant of odd intimacy between two strangers, a whole distance collapsed. My own long, harrowing journey to get there—all the switchback flights at indecent hours, the endless sessions politicking for documents from semi-legitimate ministries, dodging the surveillance of curious nearby nations and bidding for gentleness from militias, the disguises and checkpoints and satellite phone calls to editors—all of it was gone. And so was the distance traveled by the metal itself, from a weapon assembly line 9,000 miles away in the American Sunbelt to an air base in the Gulf, the pylons of a multi-role combat strike fighter, the airspace above a wind-scraped Yemeni valley where villagers were celebrating around a drill that had just struck water, and into the tissue of a man who'd never been outside his village.

I touched this man's face, all the space fell away, and what was left was just this idea: that maybe a response to the problem caused by this thing, this particular bomb, was in the story of the bomb itself.

Once I was back home, a former U.S. Navy bomb disposal technician and a Human Rights Watch investigator helped me trace the flecks of metal in the villager's eye back to a small, inexpensive weapon called "Paveway" that had been assembled by union machinists in a plant just off the airport in Tucson. But the deeper I dug, the more this one bomb revealed itself to be part of a much larger story.

BUT FOR A full year after returning from Yemen, I couldn't figure out how to tell it. And when I finally did, it was only because I happened to meet a woman from a different faraway country where this one weapon had been deployed—a woman who'd seen it threaten her own family and who then came to help my own.

She was working in eldercare by the time I met her; it was a moment when I'd begun to confront mortality at home. Parents and grandparents showing signs of frailty, a constellation of diseases in accelerated progression. My grandfather was first in line but only by a little. His health went into rapid decline, and then, before any of us quite understood what was happening, he was on his deathbed. He began to slip away on a makeshift gurney in a Florida apartment, and for a good while none of us were there.

But this woman, Radica—she was. Pop lay there threatening to die alone, and Radica affixed herself to his bedside, simply refusing to leave. She changed his diapers; she soothed and humored him. She blocked the reflexive swats when minor movements shot pulses of pain through him. She smiled at the little insults; she ignored his jokes about her accent. She brushed off the last half-assed advances of a dying flirt, and when I finally made it there, she became my friend. During those borderland minutes, watching and waiting together, she told me her story.

And after, as I thought about Radica's life—her journey from Serbia, dodging bombs in a town called Lučani, a stint in Belgrade, a transcontinental move to the strange outer planet called Florida and then to an eleventh-floor apartment on Sarasota Bay—I began to see the thing that had driven her away, this one specific weapon, in more and more unexpected places.

So the book you're about to read is in part the unlikely story of an extraordinary invention. It's partially about the way this one physical *thing* went from being a product of history to an agent changing it, a genealogy tracing aspects of the world we now live in—from technologies embedded in our daily lives to the relationships between nations—back to this one device.

But more than that, it's the story of *people*. The people who make war possible and the people who survive it. This book is an attempt to bring what happens in the backrooms and boardrooms and war rooms forward, to show its role in the blood and matter of actual human lives, in actual human minds.

The characters in each part, the real people, have never met. They are connected by a device whose purpose is disconnection. Among the unexpected things they have in common is that their lives have been changed by a weapon invented to expand the physical and emotional distance between people on either end of conflict. And, in so doing, to take people out of warfare.

This is an attempt to put them back in.

PROLOGUE

JOE

(PROJECT ANVIL)

August 12, 1944, 1815 Hours
Somewhere over Suffolk, near the North Sea

Joe Kennedy Jr. has his hands on the yoke, 2,000 feet above the English countryside in the cockpit of a rattling, war-weary plane.

Across the channel, Hitler is building a weapon. A city destroyer, a machine out of a nightmare, coming together in a bunker buried under stone and cement in Nazi-occupied France. When it's finished, it will easily reach London. Thousands will die, maybe millions, so Joe Kennedy, oldest son of a fledgling dynasty, is on a top secret and perhaps foolishly dangerous mission.

His plane is packed with nearly 22,000 pounds of British Torpex torpedo explosives. They've been crammed into every spare crack and corner: explosives packed under the nose wheel, boxes stacked up on the command deck, in the bomb bay. Hundreds of pounds of explosive bricks rattle right next to him in the pilot's compartment as he watches the countryside go by.

Joe is excited. Missions don't get any more noble than this. He's inside the world's biggest-ever bomb. He's trying to stop Hitler from destroying London.

Hitler is already launching his new inventions at England. The Nazis call them vengeance weapons, *Vergeltungswaffen*. There's the V-1, a powerful, inaccurate missile, and the V-2, based on a rocket designed by Wernher von Braun, an ambitious engineer who will go on to become an unlikely celebrity in America.

Now, for the V-3, Hitler's assigned von Braun and an army of slave labor to realize the dream: a supergun. Once up and running, the V-3 will be like a rifle fired by giants. A gun with a barrel longer than a soccer field, each bullet not really a bullet but a nearly eight-foot-long battering ram screaming across the English Channel at five times the speed of sound, crashing into London with enough energy to atomize whatever it hits.

The effect will be devastating: whole city blocks gone in a flash. And that's if Hitler fires one. The bunker Joe Kennedy is now flying toward is designed to hold fifty supergun barrels.

With underground teams reloading them, Hitler could launch something like two hundred missiles every minute. London could just disappear. And he won't stop there. Hitler wants battering rams screaming across oceans. He wants to hit New York. He wants to see cities on fire in America.

In America, generals and engineers have been scrambling to come up with a way to destroy the supergun before Hitler destroys London. Hitler's men are invisible, underground, but at battalion strength, 5,000 men working twenty-four hours a day, and the Allies didn't even learn about the new weapon until construction was well underway, and only then because agents with the French underground started reporting on something awful being built, and then a photographic intelligence expert hunched over prints of the French coastline noticed something odd: a strange arrangement of tiny crow's-feet—train tracks—running along the countryside but then simply disappearing. Tracks that stopped in the middle of nowhere. Why build a railroad spur with no destination?

Unless the tracks didn't *end* but became invisible—unless they slipped underground. Something underground—it had to be a bunker—required the kinds of supplies only heavy rail could provide.

Now American intelligence officers don't know exactly how long they have before Hitler's new doomsday machine is ready to start firing.

Engineers in Florida built massive concrete structures, panhandle renditions of Nazi bunkers, and then tried to blow them up. They dispatched ideas to England, where early air raids against the bunker didn't seem to work. It was hard to be sure—pilots had to worry about antiaircraft weapons around the bunker, flying anywhere near it was dangerous, and everything was below the surface—but after the first Allied raids, the bunker showed little outward signs of destruction, and they kept losing planes and men.

Allied engineers became fantasists, their ideas flying at fission speed, but nothing seemed good enough until someone decided to try a radical plan born from reducing the problem down to its most basic form: you couldn't get enough explosive force close enough to the supergun bunker to destroy it and still return unscathed. So, what if you removed "return unscathed" from the equation?

If you didn't have to worry about getting a plane carrying bombs safely back to base, maybe you could make the plane *into* the bomb. A kind of . . . steerable bomb.

And maybe there was even a way to control the plane—the bomb—with no one in it, using a novel technology most people hadn't encountered yet: television.

Engineers came up with a rig of rods and levers to hold the cockpit controls, and a way to control them remotely so the plane could be flown without anyone on board. They started working on a way to transmit moving images between two planes so the one they began calling the "mother" could see what the one they called the "baby" was seeing.

Early tests were promising, but there were problems that refused to yield even to the most ardent minds: they couldn't come up with a good way to get the camera mounted on the baby to move independently, which meant that, even with live images, flying it remotely would be like trying to run with tunnel vision: eyes locked forward, unable to look up or down or to the side. They'd just have to deal with that. The more pressing problem was that even once they'd figured out how to fly a plane without anyone inside it, taking off was still too complicated to do by remote control. The soft-touch coordination required to get a plane from the ground into the air was too subtle a maneuver for a few robot arms bolted to the yoke. Time was of the essence, though, so engineers and mission planners agreed on an inelegant solution: they'd put a pilot and copilot inside the baby to take off and, once airborne, the mother would take over control and the two-man crew would parachute out.

They'd just have to hope they could find pilots who, for whatever reason, would be willing to sit inside a flying bomb.

JOE KENNEDY JR. was a young man who looked out on the world and saw things he was capable of changing. His life had unfolded in a way that gave him reason to believe he was protected by higher authority—God, his dad, maybe both—and fated for great things. Confirmed as a hero as far back as the summer of '28 in Hyannis Port, when an adolescent Joe and his malnourished-looking kid brother John entered a race in their new sailboat, won, went home, then saw a boy struggling in the bay and went back out to save him, and got written up in the newspaper.

Back then, Joe had no reason to believe he was incapable of heroics, and when he began telling college friends he'd be president one day—the country's first Catholic president no less—not everyone laughed. There was no obvious reason a person like Joe Kennedy shouldn't apply his gifts, financial and physical ones, genetic ones, to the enterprise of heroism or at least leadership. People seemed to lean in to him. Their explanation was usually his eyes—striking, smiling blue eyes, the open inquiring kind with a little bit of humor in them—and sometimes just status. He'd gone to Harvard with a spending allowance the size of a financier's salary and doled it out generously for parties and sometimes friends, and he spent a college spring break road-tripping across Europe in a Chrysler, a college student not just barhopping across hostels and passing out on the cobblestones of some historic Olde Alley but taking meetings his father set with ambassadors, functionaries, the pope. He had the confidence good bone structure and money can buy. Joe, the oldest of eight, was always strapping and healthy, comfortably above John, who, as a kid, was so sick so often that his life seemed to consist mainly of recuperation and writing whiny letters about his older brother. Joe looked after Rose, the sister who confounded their parents and whom teachers called "deficient." He took her as his date to college dances, to fancy events at the yacht club, to sceney places in Manhattan.

He saw things he could make better, happier; things he could correct. A sister, a party, perhaps even a family legacy: his father—banker, billionaire, Hollywood mogul, and the last ambassador to England—had tried to stop America from entering into the war. Joe Senior had championed peace rather than a war that threatened his wealth, but now Joe Senior was home and British civilians were being killed. England was under a bombardment that would surely become much worse once Hitler unleashed his new supergun.

So, if Joe Junior was a caretaker of family, of pretty women, of country, then why not of London too? Why not also be a restorer of his father's legacy? And if there was glory to be had, a good installment in a presidential biography at some point in the future, that was only a bonus.

Plus, privately, there was something else.

Joe was the golden child and leader of the Kennedy clan, but he'd actually been having a bad year, one of his firsts. In this one enterprise, war, he'd begun to feel something awful and new. For the first time, when he looked at his peers, he was not better, kinder, more accomplished than they were. It turned out he was not a natural pilot. For perhaps the first time in his life, at almost thirty years old, he was not all that good at what he was doing, and others were catching up. Even his sickly little brother John was catching up. John shouldn't even have medically qualified to enter the war but, perhaps with their father's hand on the scale, he did. Not only that, John was commissioned as an officer almost immediately. Joe was still a cadet, suddenly outranked by his twerpy younger brother.

Joe started to struggle. He was anxious. He lost weight. He felt a new sense of urgency, a need to do more. He got sloppy. He couldn't remember instructions; he couldn't retain information. The closest he got to combat was long patrols around Puerto Rico, looking for German

U-boats that were never there because why the hell would German U-boats be near Puerto Rico? He was about as far from the action as it was possible to be while wearing a uniform, all while John went to real war in the South Pacific. As Joe stumbled further and further from glory, his brother made headlines rescuing men on an ill-fated torpedo boat mission, and it was almost more than Joe could bear. So, when his squadron was called in for a meeting and the CO asked if anyone wanted to volunteer for a top secret and extraordinarily dangerous but very important mission operating out of an airfield in the eastern part of England, it was perhaps inevitable that Joe would jump up before the briefing was even over and wave both his hands.

WHEN HE FIRST climbs aboard to inspect the modified B-17 Flying Fortress, it feels like the hold of an old pirate ship. A hulking fuselage, empty of almost everything, a giant echoing cavern with a lone pilot's seat. The copilot will have to crouch.

It doesn't feel like something that can fly. Several tons of gear have been stripped away. Every instrument, every bolt and screw not absolutely necessary for flight has been removed.

The mounted machine guns have been replaced with painted black sticks.

It's unnerving inside, but it's also impossible to ignore the sense of gravity. He's standing inside the biggest bomb ever. Once its new cargo is fully loaded—374 boxes of torpedo explosives packed around him so tightly it will take perfect choreography to parachute out alive—the baby will be the most explosive weapon in military history, a record it will hold for a year, in secret, until an atomic bomb is dropped on Hiroshima.

As he waits for the mission to proceed—practicing the parachute maneuver, supplying his squadron with good bourbon and eggs from the important people he knows at the embassy in London—the war enters a horrible final spasm. The supergun across the channel is surely nearing completion, but in the meantime Hitler unleashes his other weapons on

London. Spotters take to the roofs as they did during the first Blitz, and down in the alleys and mews and courts of London people are traumatized all over again. London becomes a three-tiered city. Communities above, people hanging on to parapets, straining to see beyond the fog; communities gathering underground and in the normal places too.

Then D-Day, the Allies landing at Normandy, Patton's army progressing. Hitler's missile tantrum: V-1s and V-2s pounding London. He's been at it for weeks, the city cowering. Nurses and American GIs down in the Tube tunnels hug each other, and they have no idea what they're in for once the V-3 supergun starts firing.

IT'S EARLY EVENING when Joe Kennedy climbs aboard the baby for the mission.

The weather is fine, a shin-high fog from the morning having burned off. With a parachute on his back, a backup strapped to his chest, and scribbled notes from the preflight briefing on his kneeboard, he wriggles up past the landing gear. He heaves himself up into the B-17's nose, and his copilot climbs up soon after.

A radio control tech is already inside, working through the last of the prep.

Joe settles into the pilot's seat. He puts his hands on the controls and looks out the left side window, where a mechanic standing on the tarmac gives him a thumbs-up.

Joe begins the engine start-up sequence. He starts with engine three; the mechanics said engine three gave some trouble during morning check, not holding its RPMs, and now it protests, it screeches and whines, but soon it's chugging. The propeller starts to turn, and Joe continues the process. One by one, the 1,200-horsepower engines cough themselves awake, spit flames back, and sputter to life, all four of them beginning as a rumble, rising to a shriek that vibrates the metal shell, so the boxed explosives stacked around him go a little blurry.

He raises his hands up in front of him so out on the tarmac the ground crew can see, sticks his thumbs out, and yanks at the air. Crewmen scurry under the plane to pull the chocks from the tires.

The B-17 is free.

He nudges the throttle, engine noise rising across the airstrip as the massive warhead tries to heave itself into motion. The wheels begin to turn, the plane moving now. Joe taxis to the end of the runway, and it's time for the final test.

With the plane positioned for the takeoff run, Joe lifts his hands off the controls and waits.

There's a sudden bang from behind him. On the plane's tail, the flaps have clanked down, as if all by themselves. It's eerie. It's almost funny. From somewhere out of sight, the mother, already airborne, is manipulating his plane.

"Down elevator," he confirms over the radio. He waits some more. Another clank; a fin on his right wing moves.

"Right aileron."

Joe gives the radio tech next to him a look. The tech shakes the copilot's hand, then looks back at Joe.

"Wish I was going along with you."

"You'll make the next one," Joe says. He tries for a little gallows humor. "And if I don't come back, you can have the rest of my eggs."

The tech nods, retreats, drops down from the plane. Joe tests the engines one more time; engine three still a little off, something not quite right. He signals for a mechanic to come back for one last check. They decide it's good enough; it will have to do. The mechanic hustles back down and out of the plane's way, and Joe looks down the runway. It's time.

He puts his palms over all four throttles. He pushes them all forward, all together, all the way.

The squatting copilot grips the side of the seat as four propellers claw at the air, the weighted-down baby resisting, like it has its own opinion about the mission, and it takes a few long breaths before the wheels begin to roll again and the whole creaky thing is moving, taking too long to get to speed, eating up too much runway, staying on the ground for as long as it can and closing on the field beyond. And then finally, at the last moment, the baby concedes and lifts into the air.

FOR JOE KENNEDY, riding a guided bomb up toward the cloud cover, there's now relief. Not just because the lumbering thing is airborne but because he's been worried for weeks now that the war will be won before he has a chance to help win it. Now here he is, flying the bravest, most important mission there could be, on the way to strip Hitler of his most fearsome weapon, maybe to save thousands of people in the process.

He eases the plane east, toward Nazi-occupied France, and rendezvouses with an escort formation: small, nimble fighter planes that will fly part of the way alongside him. And since this is a war in which everyone who can serve is serving, Joe isn't the only royalty on the mission: just off his wing, Elliott Roosevelt, son of the sitting president, joins the formation in a de Havilland DH.98 Mosquito, a lightweight plane made of wood, fast and agile and packed with cameras to capture the final moments of Joe's mission when he parachutes out of the baby, down to safety, to begin his march toward his planned future. Congressman, senator, one day president of the United States.

Joe draws closer to the coast, still low enough that he can see the individual conifers that make up the forest through his cockpit window; low enough to register the surprise of colors on the heathland, moor grass so rich it looks unnatural, like purple carpeting beneath the trees. He's almost low enough to see the two young boys who will later report that they looked up when they heard the rolling, growing chug and saw a spreading formation unlike anything they'd ever seen before, even in war. A strange flying parade with different kinds of planes—dozens of small fighters, a wooden one, a few bombers—all of them passing low over the countryside, enough hardware up there to shade the trees.

Joe levels the plane at 2,000 feet and prepares for the last steps: setting gyroscopes and warning lights, turning on the autopilot so the mother can take over.

He's now just a little off course but mostly on schedule. He clicks the radio and says two words: "Spade Flush." A throat mic picks up the pulse of his vocal cords; rubber sensors against his larynx convert his voice to an electronic signal that is then transmitted across the air to the mother flying behind him, where pilots hear a version of Joe Kennedy in their headsets.

In the mother, the pilots confirm with each other that they've just been given the code phrase and they take control of Joe's plane.

Now Joe is sitting in a vessel he no longer controls. He watches, hands off the yoke, as his plane starts to turn.

Joe and his copilot stay there for a moment, just riding. They fly for another ten minutes, making sure the connection between the two planes remains stable as the mother steers them toward a series of checkpoints and Joe prepares to bail out.

The jump from the plane is what's most frightening. At two hundred miles per hour, the slipstream can grab a body and slam it back up against the plane's underbelly. Even if he clears the plane, when a chute deploys at that speed, the force on the harnesses is so sudden and so powerful, the sensation will be akin to having both his legs severed. Joe Kennedy is, in effect, about to jump into a hurricane.

But he's prepared. As prepared as can be; at least he knows what to expect.

Twenty minutes in, the secret mission is still proceeding mostly according to plan. They're a little off course, but there's enough time to get back on the assigned route, planned and timed around lighting conditions for a strike from due west at 1900 hours, when the supergun site will be illuminated for the camera and any gunners on the ground will be looking right into the sun—when the effect will be of the bomb flying out of the sunset.

The mother steers Joe west.

Roosevelt eases his wooden plane closer, into position to photograph Joe and the copilot leaping out.

Joe's work is almost done. He pulls a safety pin to arm the explosives.

He prepares for the jump. All that's left to do now is power on the television camera so the baby can keep flying past the mother plane's line of sight while the mother stays a safe distance away from the supergun bunker.

Joe's copilot throws the camera switch.

In the mother plane, a television monitor flickers on, and for just a moment the crew can see what Joe sees.

What they can't see is a series of small flaws in the onboard electrical system that, during testing, caused the arming panel to keep lighting up on its own as if receiving some phantom transmission. Most of the engineers dismissed it as the result of random ambient FM radio signals, but a few continued to worry that it might have something to do with the television camera mounted on the front of the plane. Once powered on, the camera emits energy, which can potentially throw a nearby electrical switch that isn't well shielded.

By a coincidence of improvised design, there happens to be a switch housed right next to the nose camera. The switch that's connected to the detonator, which, in turn, is connected to 22,000 pounds of explosive.

JOE'S PLANE DROPS from radar displays as a sudden wind rocks the other planes in the formation up and then down. On the monitor in the mother plane showing the camera feed, the English countryside turns to fuzz, and from a garden below the two boys watching the strange parade of planes cannot understand what they're seeing. Inside the formation above them, one of the planes has transformed into an expanding disc of flame, pierced through by a giant orange beam flying up into the sky and down to the forest floor. The aircraft has been blasted into a thousand pieces by an explosion so powerful the shock wave launches doors off hinges and drives ceilings down into houses hundreds of yards away. The two boys watch as thirty tons of superheated steel and aluminum scream across the countryside, cutting through trees and turning the ground to hellfire.

BOOK I

HILTON

(OPERATION ROLLING THUNDER)

1.

Twenty-three years later
Somewhere over North Vietnam

Lieutenant Colonel Richard Hilton rockets through the sky at close to the speed of sound, five miles above enemy territory. His plane bobs in its formation. A small ship at sea, vortices of wind across the skin nudging the airframe up a little, down a little, a little over. He's at .88 Mach, a good speed, a good, smart speed, he thinks, not cooking through fuel but fast enough to fight if the enemy jets engage.

Which they almost certainly will.

Rick Hilton is a savant in the cockpit. Fused to the machine and sealed off from emotion, capable of rapid calculation at speed. He processes extraordinary amounts of information almost instantaneously. *Pull up, bank, engage afterburners*, he turns data to decisions and decisions to action in the time it takes for a ripple of electricity to flash across his brain.

He's at home up here. All he ever wanted to do was be in the sky, ever since he was a mouthy kid in Oklahoma on a Sunday drive and caught the sun glinting off a dinky old crop duster flying alongside the family sedan. A four-year-old Rick Hilton blurting out to scandalized parents, *look at that sum' bitch go!* Then, after his parents' marriage blew up, a stint being passed from home to home and idle time spent at the local airstrip waxing single-engine props for the chance to ride in them, then teenage years getting into the air however he could, even when it just meant finding a ramp to ride a heavy old Harley off of and get away from terra firma for a breath or two.

Now he's up in the thin air, in a real live war. Up above the matted clouds because a supply line snakes through the countryside right below and the enemy's protecting it with everything they have. Destroy that supply line and save Americans. Leave it intact and the enemy keeps feeding a beast chewing through the kids not lucky enough to do Vietnam the way he's doing it—the kids without 20/15 eyesight and the sealed-off kind of mind for math at altitude Hilton has.

The supply line is the problem, but as far as Hilton's concerned, the whole war might as well be about one especially brutal section of it, a 540-foot span of steel and rebar known as the Dragon's Jaw. The only railroad in the panhandle crosses that bridge. It's the only way the enemy can get supplies across the Sông Mã River, so the enemy has it defended better than any target in the country, which right now means it's defended better than just about any target in the world. Hilton's seen just about every kind of mission launched against that bridge, and they've barely dented it.

If Hilton survives this mission, it'll be the Dragon's Jaw that all the officers and pilots and even the Thai laborers back at base who wander across the tarmac to find work will talk about. They've lost friends over it. Enough failed missions, enough planes shot through and lost in the area around the Dragon's Jaw, that whole mythologies have risen up around it: that the bridge is the stitch that holds the universe together; that it's a link between two dimensions; that it doesn't exist on the terrestrial plane at all. A trick of the light—some kind of hologram drawing pilots in like mosquitoes to porch lamps.

For Hilton, it's only by dumb luck that, for all the deadly flying he's done over the supply line, he hasn't yet been tapped for a mission against the Dragon. But he knows. That luck won't last forever. If he survives today, tomorrow it'll be the Dragon, and without an effective weapon against small targets in dangerous airspace he's no more likely than all the pilots who've gone before to make it out alive.

2.

Okaloosa County, Florida
Air Proving Ground Center
Eglin Air Force Base
"Detachment 5"

To Weldon Word, a blue-eyed thirty-four-year-old engineer, the problem was obvious. And almost instantly, he began to see how he might solve it.

Just called up by his employer, Texas Instruments, from a dead-end post off the coast of Rhode Island—a year and a half sitting on ships scraping up miles of sensor data for a sonar program theoretically useful to someone but so monotonous and incremental it was hard to feel any kind of impact—Weldon had been waiting for an escape, primed to dive into a problem with real stakes, when he got the call from TI dispatching him down to Florida for an important new project just now coming into focus.

Across from him an overenergized colonel paused a monologue about the dangers facing American pilots, pulled a file from a desk drawer, and slid it across the table to Weldon: an aerial reconnaissance shot in black-and-white. An important bridge, the colonel explained, over in Vietnam.

To Weldon, it looked like a bridge on the moon. From bombs, the colonel said—all bombs that missed the target, and this didn't even account for all the bombs that missed into the water. Weldon counted: ten, twenty, one hundred, two hundred . . . there had to be eight hundred craters near the bridge and, from the looks of it, almost no damage at all to the bridge itself.

Weldon looked up from the photo and trained his blue eyes on the colonel, already back at top speed, holding forth on his desire to provide the boys flying in Vietnam with a new kind of weapon, one that would at least give them a chance to land a few blows against this damned bridge before they got themselves shot out of the sky.

The colonel shifted to his wish list and began rattling off specifications.

The new weapon would have to be effective enough that pilots wouldn't need to linger in dangerous airspace for long or return to it again and again—a weapon that could help do away with massive "alpha" strikes and "dual axis" attacks that put hundreds of pilots in the line of fire.

He wanted something that might allow just a few pilots to do the job that right now required dozens to even attempt.

He wanted a weapon that could be deployed from farther away. From, say, 10,000 feet or so, high enough to keep the boys up there a little safer a little longer. And as Weldon listened—looking from the moonscape recon photo to the colonel and back, imagining how such a weapon might work—three disparate threads started to converge in his head.

First, he was thinking about a new technology in the works, something Albert Einstein had dreamed up and the comic books seized on. Einstein had an idea that, at least theoretically, you could use radiation to amplify a beam of light to perhaps as much as a million times the intensity of the sun. To the United States Army, that idea—light amplification by stimulated emissions of radiation, or LASER—had at least one obvious application: a death ray. Over time the idea evolved into a beam that maybe didn't kill but would still maim an enemy soldier, who would then need rescuing and medical care, removing not one but several fighters from the battlefield. The Army contracted with a company called Martin Marietta to work on LASER, but it turned out generating a beam that could harm an enemy took more energy than could feasibly be packed into any kind of remotely mobile package. Martin Marietta managed to build prototypes, and the laser beams could reach long distances, but not with enough power to do much to an enemy. All they could really do was, essentially, point at things. Even then, not all that usefully, since the versions that actually worked operated on a light frequency too high for the human eye to see. The Army had considered using LASER to try to guide one of its ground-launched missiles, but Army medical officers got in the way. LASER wasn't powerful enough to disable a vehicle or injure an enemy soldier on the other side of a battlefield, but it was plenty powerful to blind an Army missile crewman standing beside it.

By the time Weldon landed in Florida, the sci-fi promise of LASER had given way to the more sober reality of what was essentially just a long-range flashlight with an invisible beam, more dangerous to the user than the enemy, and lacking any practical use. But, sitting there in the colonel's office, it occurred to Weldon that he might have found one.

If the Army's interest in finding some way to guide a ground-launched weapon with a laser didn't make sense because the person operating it might accidentally look into the invisible beam, that, theoretically, wouldn't be a problem with an air-launched weapon, since there was little risk of a pilot in the sky accidentally looking into a laser beam pointed at the ground.

When the colonel paused for a breath, Weldon tested the water. "Well," he said. "How about laser-guided stuff?"

IF THE LASER was the first of the threads coming together in Weldon's head, it had to do with a story beginning fifteen years earlier with the Nazis and a man Weldon had come to know well: Tom Weaver, who returned from World War II having made friends of his enemies.

No one knew the whole Tom Weaver story, and he seemed to like it that way. He'd lost a leg; no one knew exactly how, and he gave a different explanation every time someone asked. The one thing Weldon knew about Tom Weaver was that he'd been involved in Operation Paperclip, the top secret mission to bring scientists from the defeated Nazi regime to America. Whatever Tom's job during the war had been, his job after it had been working with a team of Germans relocated from the Reich to Alabama, where they were given U.S. government housing at the Redstone Arsenal near Huntsville and a generous budget courtesy of the American taxpayer and put to work repurposing Hitler's rockets for the United States. Tom worked with the men who designed the "vengeance" weapons Hitler used to terrorize Europe—the very weapons that inspired Project Anvil and drove the Allies to dream up the remote-control B-17, the ill-fated early entry in the quest for a steerable bomb Weldon now found himself in the middle of.

But from early on, the relocated Germans showed a key weakness. They'd proven to be perhaps the best in the world at launching things but were still not very good at controlling where those things came down. Steering hadn't been a high priority when they worked in Germany: Hitler had wanted to terrorize cities full of civilians, and it hadn't much mattered which particular building was destroyed or even what neighborhood a rocket landed in. Hitler's V-2 could travel a hundred miles or more, but it was barely aimable.

To turn Hitler's V-2 rocket into something the United States could use, it needed to follow a path. To follow a path, it needed to know when it was falling off course. The Germans had access to instruments that could do that—to sensors that could generate electronic currents—but for an electronic current to really *do* anything—to move a fin, for example—that current would need to be more powerful. It would need to be amplified. At the time, the only available "signal amplifiers" were heavy glass contraptions, cylinders that needed to be vacuumed free of air and sealed off so oxygen molecules didn't get in the way of electrons, and which could then turn a small electronic signal into a more powerful one and pass it on. The signal-amplifying "vacuum tube" was so key to so many early electronic devices that those devices were often given nicknames derived from it: "tube" for television; "amp" for the box that turned the nearly silent vibrating strings on an electric guitar into sound coming out of a speaker. And on a flying rocket, "vacuum tubes" were how you turned the signal from a sensor saying "time to turn" into a current strong enough to move a fin.

Which meant that every time the German scientists launched a test rocket, a few dozen heavy glass tubes went up with it. To turn an unaimable weapon for a revenge-blinded Führer into an actually useful rocket, the scientists were going to need a lot of practice, and for that they'd need a generous supply of those vacuum tubes.

So Tom Weaver hobbled out of Alabama and went hunting for a way to supply the underperforming German scientists with more vacuum tubes—or, perhaps, with some kind of substitute.

THAT WAS THE second of the three threads beginning to come together in Weldon's head as he sat across from an angry colonel with a picture of the bridge between them: strange old Tom Weaver and his one-legged quest to help aim Nazi ingenuity at America's enemies. The last one had to do with Weldon's employer; with the very founding of the company, and a corporate origin story that had led to this very meeting with an impatient officer of the United States military.

The company that would become Texas Instruments was incorporated on December 6th, 1941, to sell a promising new product for the oil prospecting industry. The very next day, the Imperial Japanese Army attacked Pearl Harbor, prompting the United States to join the war and take control of oil exploration, which meant the first strategic decision they ever made was a giant bet on an industry that looked like it might be disappearing overnight.

What the small team of brainy oilmen lacked in timing, though, they tried to make up for in creativity. The invention they'd intended to sell was a "magnetic anomaly detector," a device designed to pick up inconsistencies in the earth's magnetic field that suggested the presence of hydrocarbons. But if looking for subsurface magnetic anomalies to find oil had become a profitless endeavor, they figured there might now be a different subsurface magnetic anomaly worth looking for: enemy submarines. They hooked up with the U.S. Navy to work on a better method for detecting German U-boats, eventually changed their name from the generic "Geophysical Service Incorporated" to the even more generic "Texas Instruments," and secured enough work with the Navy to survive their early misfire, though the embarrassing first step served a complex the company never quite shook: a preoccupation with trying to avoid obsolescence. If you couldn't count on today's customer, try to anticipate tomorrow's.

Some of the naval officers TI had worked with during the war became the company's more influential executives afterward, chief among them a young reservist and budding prophet of technology named Pat Haggerty, who said that the war had given him a glimpse into the future.

Pat Haggerty came out of the war preaching to anyone who would listen that the course of the war had been determined by technology—specifically, electronics—as much as it had been by strategy. In Haggerty's telling, the Japanese had an advantage early on, one he swore you could attribute to factors as simple as their better binoculars and night vision. Once the Allies started relying more on radar and sonar to find Axis ships and planes, the advantage swung to the Allies. Or, Pat said, look at the planes. For a while the German Messerschmitt and the Japanese Mitsubishi Zero were nearly invulnerable—superior feats of engineering that outclassed Allied pilots and antiaircraft guns. But once the Allies started using "proximity fuzes," enemy planes were no longer immortal. Close enough was suddenly good enough, and the course of the air war changed.

Pat's vision began to extend beyond the military, and he wasn't shy about proselytizing to anyone who would listen, Weldon included. The same electronics that had given the U.S. Navy an advantage, Pat said, could one day be useful to American consumers too.

One particular technology he thought Texas Instruments ought to take a look at was a curious new device called the "transistor." It was a hybrid of two electronic phenomena. Everyone knew what an insulator was: a material that blocked electrical current. And a conductor was a material that passed an electrical current on. A transistor, though, could do both. The transistor was a sometimes conductor, a *semi*-conductor. And if you could build a device that could stop a current but also pass a current on, Pat reasoned, what you'd have, in essence, was something that could make simple decisions.

The telecom giant Bell Labs held the fundamental patent for this sometimes-conducting device, but the U.S. government was worried about a monopoly and forced Bell to license it to other companies. It still wasn't clear exactly how big a market there might be or even what the best uses for such devices were, but when Pat began urging Texas Instruments to buy into the idea, he was trying to persuade a corporation already susceptible to a pitch for a product that would serve a future market rather than the current one.

By the time Weldon joined the company, TI was pursuing the "semiconducting" device with the zeal of believers atoning for original sin and doing it in the most difficult but, they believed, most promising way. The work on semiconductor technology thus far mostly used an element called germanium, which was expensive, scarce, and stopped working at high temperatures. Germanium was generally agreed to be less than ideal, but the element theoretically best suited for the job was also the trickiest to work with; nonetheless, TI decided they might as well try. They took their license and committed to spending whatever it took to figure out how to make a sometimes conducting device out of the more temperamental element, finicky and frustrating and more brittle than germanium but almost as abundant as oxygen: silicon.

Trying to manipulate a substance as fickle as silicon presented a host of problems. Silicon was abundant in nature but silicon pure enough for TI's purposes was still almost nonexistent. They needed to find a way to grow silicon in crystal form, then purify it even further; and even *then*, silicon was highly reactive: it absorbed impurities from whatever was around it. TI needed to find a way of working with it while keeping the facility pristine; they had to avoid contamination; they needed special instruments. But TI was determined to be the ones who cracked the code, solved the silicon problem, and emerged as the leaders in . . . they still weren't entirely sure what.

Even Pat Haggerty, emerging oracle of future technology, couldn't foresee that a silicon semiconducting device might become to the twentieth century what the steam engine had been to the industrial revolution. The early versions were unimpressive-looking things, rigid little decision-making devices—*chips*—but none of that mattered. It didn't matter that they couldn't possibly know what the future held. They just wanted to be the ones driving it this time, the ones *causing* the future, not just responding to it. TI wanted to be agents of change rather than victims of it, and they figured it wasn't a failure of imagination that they couldn't yet see how the chip might be used down the road, but rather a reflection of its awesome promise.

Still, TI knew that the future wasn't guaranteed. Even if they solved the engineering problems and materials sciences problems, there was still the economic problem: before they could realize the dream of electronic decision-making devices finding their way into consumer products, the chips had to get much, much cheaper. And to make them cheaper, Texas Instruments needed to find a way to sell a lot of them. They needed to find a customer who wouldn't balk at a very high first-mover cost on a mostly untested product—a customer who, for whatever reason, would be willing to buy TI's new chips like they were going out of style.

And it was around then that a one-legged storyteller named Tom Weaver started popping up at electronics conferences with a generous budget courtesy of the American taxpayer and a pressing need to replace the dozens of vacuum tubes burning up every time his ex-Nazi scientists launched one of their errant test rockets.

BY THE TIME Weldon joined Texas Instruments in the late 1950s, there was already a sense of destiny around the unlikely marriage between the team of ex-Nazi scientists Tom Weaver was working with and TI's Pat Haggerty–led push into chips.

With the TI chip replacing the more cumbersome vacuum tubes, the Redstone rocket program entered a period of rapid improvement. The rockets got better almost overnight: Hitler's vengeance weapons evolved into the Saturn V rocket, which would eventually take the Apollo mission to the moon, and the Minuteman ballistic missile, capable of delivering a nuclear payload from the Midwest to Siberia. The Redstone program became so impressive so quickly, it captured the attention of leaders even in industries beyond technology. Walt Disney took notice. No stranger to spectacle himself, America's foremost entertainer dreamed up a way to borrow some of the Redstone program's sheen to market a project he was working on. Disney had a parcel of land in California where he planned to build a giant tourist attraction. Since part of the plan was a "themed land" called "Tomorrowland," designed to awe visitors with a glimpse of the future, Disney called on Redstone's lead engineer, a handsome former Nazi officer named Wernher von Braun, to see if he might serve as a kind of spokesman. Von Braun was just a decade removed from his service to Hitler in the SS, but Disney saw promise, and together they made a series of television specials in which von Braun, addressing the camera with his German-accented English, taught the public about the future of space exploration and, while he was at it, about the exciting new theme park opening out west.

Von Braun proved to be a natural. Tom Weaver's lead engineer charmed American audiences, who didn't seem to know or care about von Braun's past. The first special was beamed into 42 million American homes, making it one of the most widely viewed television programs in the young medium's history, and turning von Braun into a household name. *Time* and *Life* magazines both put him on their covers, and the former Nazi officer found himself mingling with movie stars, business leaders, titans of industry, and politicians—including one first-term senator with whom he shared an uncomfortable connection. After tracking von Braun's rise to prominence and looking into the charismatic engineer's past, Senator John F. Kennedy requested an audience. When the two first met, at a film studio in Manhattan, Kennedy showed a peculiar interest in one particular weapon von Braun had developed for Hitler, and once they exchanged pleasantries, Kennedy let on the reason for his interest. His older brother Joe, he explained, had died trying to destroy the weapons von Braun had built while flying inside what had been in effect a giant steerable bomb. Nonetheless, the two emerged from the charged first meeting on congenial terms, and von Braun would soon count Kennedy among his friends in the American elite. When Kennedy was elected president, he'd turn von Braun into perhaps the single most important figure in the race to the moon.

Texas Instruments, meanwhile, had stumbled into a key realization while playing its background role in the rising spectacle of America's rocket program. Tom Weaver's early purchase orders proved to be exactly the spark the company needed for its new product category. Supplying the ex-Nazis in bulk was beginning to bring the per-chip manufacturing cost down, and saw that if it really wanted to sell enough chips to drive the cost down to where it could spark a new market, they ought to look for more clients like the ex-Nazis working on rockets. TI had inadvertently been handed a road map to product viability by showing them there actually were customers unbothered by the high price of a new product. Government clients. Defense clients. And if there was a drawback to depending on rockets to spark a market, it was only that rockets were big and complicated, and even in a period of rapid experimentation there were simply not that many being built. The rocket program had proven to be a big, generous customer for TI, but didn't represent an unlimited market.

So TI executives started to look for defense customers who, like Tom and the ex-Nazis, might be willing to pay high prices for new chips but who, unlike Tom and the ex-Nazis, might be interested in experimenting with weapons that weren't massive and awesome but smaller and cheaper and, therefore, potentially more plentiful. But Texas Instruments still had hardly any record as a defense company, its contribution to the Redstone program critical but unspectacular to those who even knew TI had a role in it. The company gambled that it might be able to overcome that by leveraging a developing trend in defense contracting that looked like it just might provide an opening for new players in the defense space.

As American involvement in Vietnam accelerated, the U.S. military was finding itself ill-equipped for the type of war it was fighting, and since defense contracting was arduous and bureaucratic, the prospects of getting better-suited equipment to warfighters with any kind of speed were dim. The Air Force was especially keen to find a way of getting new and better weapons to the theater, both because the brass prided themselves on the Air Force being the most innovative branch and because what the pilots had was so clearly inadequate and they were getting shot down because of it.

So, quietly, the Air Force came up with the idea of making weapons development faster by making the scale smaller. They set up an outfit called Project 1559 at Air Force headquarters in Ohio with a small detachment down at the Florida base where the Air Force did much of its testing. They assigned a hard-charging colonel to oversee it, gave him a small budget and a large remit to take small bets on new ideas, and it was there, at the Project 1559 office in Florida—where the colonel took his assignment and focused it almost exclusively on the one undroppable bridge killing so many American pilots—that TI figured it might have a chance. It was still a gamble for the company, though, so, rather than throwing their best men at the project, they plucked one promising young engineer from a mostly meaningless project in Rhode Island and sent him down to Florida to help a company still unproven as a defense contractor get its "chip" into more weapons.

Or, failing that, to perhaps invent a new one.

3.

Somewhere over North Vietnam

Rick Hilton listens: even slow moments in combat aren't quiet. The helmet helps to muffle the noise, padding pressed up against his face and squeezing his cheeks to freeze his jaw in place so the bones don't conduct sound. The howl of jet engines a few feet from his ears, the shriek of wind scraping the skin of the airframe and heating it up, warming the cockpit so he's sweating even though it's thirty below zero outside—it's all there, all constant, all lowered down to bearable by a helmet that grasps his head like a pair of downturned hands, guiding him toward what he's supposed to know.

He looks toward the ground, but a cloud deck blocks his view. A solid gray carpet cutting him off from whatever's below. He knows MiGs can pop out without warning, the Russian-made fighter jets smaller and faster than his heavy old Phantom. They can get position on him with their two cannons, and when those things get to spinning, they can haul off as much as a thousand rounds a minute, almost a thousand pounds of steel coming at him.

He needs to see what's below those clouds. He needs to be lower.

This whole formation of planes should nose down and descend *now*, get below that cloud cover so they can see the ground before something streaks up from below. And if he were in charge . . .

But he's not in charge.

Patriot, proud American, Kennedy supporter in the process of converting now that he's gotten himself into a war he's not really allowed to fight.

He has no cannons. He has no guns at all. All he has to defend himself are a couple of finicky missiles that are supposed to track an enemy plane using some kind of unperfected radar technology, but they fail more than they work. When MiGs show up he'll be virtually defenseless. He doesn't have a weapon effective for the kind of war he now finds himself in. He doesn't have weapons for any of the situations he finds himself in, that the politicians have put him in.

It's Washington's fault. It would be far less dangerous to attack enemy MiGs when they're on the ground and unmoving, but American bombers don't have a weapon that can reliably strike a precise target, so they can't destroy planes parked on the ground without dropping a whole train of bombs. Which means a chance of hitting things nearby, maybe a delegate from Russia or China there to advise the North Vietnamese military. For the secretary of defense or the president or whatever feckless senator it was making the calls, the risk of provoking Brezhnev or Mao into greater involvement in Vietnam was apparently too big to accept; much better to just risk American pilots. The only way they'd let American pilots over Vietnam attack the MiGs defending the Dragon's Jaw was to bait the MiGs up into the air. And when enemy jets are up in the air, they can shoot back.

The radio is bothering him.

Sometimes the voices come from other planes nearby; mostly they come from planes behind him, above him, over the horizon somewhere, from planes he can't even see. Staticky voices rise above the hum, piped right into his head like he's imagining them. He's noticed that when men at altitude get nervous, they start to chirp at each other—some urge to confirm they're not up there alone.

Hilton doesn't like it. There are too many things to pay attention to. Information blinks up at him from an instrument array with a dozen knobs and dials, a strobing glareshield in front of him with its own vigorous appeals for attention. And in the rare seconds when there's no sound coming from other places, there's still Mike, in charge of weapons and riding in the back seat, on his first mission. Latched onto one another by intercom, linked by breath. It's intimate, almost sensual; in those moments when nothing else is reaching out to shriek in his ear, Rick can sometimes tell how Mike is doing just by listening.

The radio crackles.

"This is Red Crown on guard." The Navy ship. "Red bandits Gia Lâm, Phúc Yên."

So: enemy jets have just taken off from two nearby airfields.

They know we're here. "Red Crown, on guard." The Navy has more. "Red bandits Bullseye 030 for five." "Bullseye" is code for Hanoi; the MiGs are five miles from Hanoi on a northwest bearing. "Red Crown, Bandits 030 for 15."

They're turning harder toward the north. Hilton puts the heel of his hand on the throttles and eases them forward. The airframe vibrates, his chest squeezes back into the seat as the jet pulls toward supersonic, and he gets ready to fight.

ON THE GROUND, a North Vietnamese technician is sitting in a cramped van cabin, watching a wall of monitors, each screen showing a picture of the airspace from radio waves sent and received by the giant rotating "acquisition" radar on the van's roof, when Hilton's plane blips into view. A dot on one screen shows Hilton's altitude, a dot on the other shows his distance and heading.

The technician quickly jots the coordinates and radios them to a different van with a different kind of radar on its roof. Inside the second van, members of the crew stand from their own walls of screens and dials, then reach up and manually turn a cabin mount above them, rotating and tilting an engagement radar array so it faces the coordinates they've just received—a whole arrangement of vehicles and equipment and a few dozen people all required to find and engage their target. It's big and expensive and unwieldy but mostly sufficient for its purpose.

They set the second radar to SEARCH mode, watch the monitors, and within seconds Hilton's plane shows up on their screens too. Target acquired. The engagement radar crew tightens the net. A crew member switches the radar from SEARCH to a TRACKING mode.

Above them, the engagement radar begins emitting shorter, more frequent pulses toward Hilton.

HILTON STILL HASN'T even seen the enemy jets before Red Crown comes back on the radio to say the MiGs crossed into China and are gone. That was bizarre, Hilton thinks. Enemy jets took off like they meant to engage but then flew right out of the fight. A riddle to solve, and he's working the problem in his head when the cockpit erupts in a riot of warning lights and the helmet speakers start screaming in his ears, then change to a panicked rattling, and at the exact moment Hilton realizes what has just happened—that the enemy jets have led him into a trap—voices from men in the planes nearby explode over the radio.

"SAM! SAM! SAM!"

The Navy ship out in the gulf joins the chorus in his ear.

"SAMs! SAMs!"

On the ground the engagement radar team watches Hilton's plane become a formation of planes, multiplying dots on the screens. They radio a third crew, this one manning an arrangement of missile launchers hidden in the trees near the radar vans. The missile crew rips camouflage tarps off the launchers and runs to the control panels; the missiles begin to swivel and rise from their hidden positions while a technician in the engagement radar van watches the monitors, waits until the planes have passed just overhead, and issues a Fire command.

Outside, a solid-fuel rocket engine ignites, a contrail blasts out the back, and a thirty-five-foot weapon leaps off the rails as if sucked into the sky.

Up in the cockpit, Hilton's plane screams more warnings and lights flash across the glareshield. Within six seconds the missile behind him reaches the speed of sound. Its first stage falls off and the liquid fuel engine ignites, the weapon glowing angry red and doubling his own speed, then tripling it, covering a distance of miles in a matter of seconds, and he knows he's dead in the water. He has no way of evading it. His only hope is that he's fallen so badly for the trap that the missile was actually fired from too close and if it's a little off target it might run out of time to correct. It races up toward him, still hidden by the cloud cover, receiving radio commands from the engagement radar van below: *Steer up. Steer up.* The missile noses up at him, still accelerating. It reaches Mach 4 and it's like Hilton is standing still.

But he's right: the SAM is catching up too fast. It receives its last instruction from the engagement radar van below—*Steer up*—and in an instant it's overtaken the plane.

Hilton sees the missile blast out of clouds in front of him, still honoring its last command from the radar van, bending up until it's almost vertical. He watches it race halfway to space before exploding and sending shrapnel at nothing in a spectacular fireworks display.

Disaster averted, but only from dumb luck and an enemy weapon's imperfect steering. Behind him, Mike sighs. Hilton unclenches.

DOWN BELOW, ANOTHER solid-fuel booster ignites and another SAM rockets into the sky. Then another and another.

The speakers in Hilton's ears scream somehow louder and louder, and it all becomes too much: the plane isn't telling him where a missile is coming from as much as screaming at him that dozens of missiles are coming from everywhere. Almost thirty SAMs are in the air now as if the earth got angry, erupted, and ejected telephone pole–size quills at four times the speed of sound. Hilton's instruments have stopped giving him information and are now only giving him noise, a cacophony of different sounds that together mean nothing, the cockpit a locked-in space of sensory chaos.

Hilton has to suppress the urge to withdraw entirely, to take off his helmet, take off his flight gloves, and throw them over the dials. Pretend he's in that old simple Oklahoma crop duster, because none of the noise is helping and none of it changes the fact that he's going to die in a few seconds anyway.

He can picture the crew below, celebrating their triumph. From the briefings he knows what it looks like: the whole splayed, cabled arrangement, a networked organism with a dozen or so vehicles. He knows how many men must be down there, all trained and motivated and focused only on getting their weapon to steer right up his ass. All that manpower and equipment versus him, one pilot, one plane, alone in this machine that has maybe a few more seconds of flight time left in its life. Alone except for Mike, breathing in the seat six feet behind him. Mike, who's on his very first mission.

Hilton decides to try. One man versus the whole team below firing missiles that fly so fast, he might as well be going in reverse. You can't outrun something that fast; the best you could do if you had the balls to try was to turn toward the missile and fly right at it. An ultra-high-speed game of chicken, just a little bit off center so you have some depth perception. Nudge the stick forward, lower your nose a little to see if the missile lowers too—to see if it's definitely *you* the missile crew down below wants. Wrap your fingers around the stick, let the missile approach, closing with a combined speed of Mach 5, wait until a half second to impact, yank the stick hard. Full roll, go inverted, let the missile fly right by your upside-down head. The timing has to be perfect. Move too early and the missile will have time to correct; too late and it won't have to. Hilton makes himself focus. He keeps his eyes locked on the clouds below so he'll see the next missile as soon as possible. He tries not to blink.

He tries to remember to breathe.

He tries to block out the screams of men in other planes, disabled and hurtling toward the earth. He squeezes his fingers around the stick, gets ready, and then, just as suddenly as it began, the barrage stops. The alarms go silent, the warning lights stop flashing, as if the earth has opened back up and recalled all its demons.

It's suddenly quiet. The clouds erupted in arcing thirty-five-foot missiles, and they've somehow managed to survive unscathed. A sigh of relief; someone must be looking out for him.

He allows himself a moment of calm. He tries to get his pulse down—he still has to navigate back to base, and there may be an issue of fuel. And as he draws his first deep breath, one last missile pierces the cloud cover, turns toward him, and accelerates. An antenna in its nose bounces radio waves off his plane until the interval between signals is short enough, which means the missile is close enough, and a proximity fuze triggers the missile's explosive payload.

4.

Rick Hilton is 25,000 feet above enemy territory when he has a sense memory. He's standing over the kitchen sink with a dish towel during a California storm. A sudden crack, and an old oak in the yard splits under a lightning bolt.

A blast wave shocks Hilton's cockpit, and a few hundred pieces of shrapnel fly through the fuselage, rip off the plane's antenna, chew through the primary flight control systems, and rupture a hydraulic line feeding the backup. Jet fuel begins spilling out into the slipstream. Hilton's radar scope goes blank and the gauges for his left engine do the opposite: they go blurry, needles spinning so fast he can't even see them.

"*You're on fire!*" someone, another pilot somewhere nearby, is yelling in his ear. "*Get out now!*"

He means eject. Behind him, Mike is likely obliging. Chin in, legs in so he won't lose a limb when the rocket beneath his seat ignites and blasts him out of the plane. Mike says he's ready.

Hilton hears himself saying, "Don't go yet."

Almost without realizing it, he's begun working on another option. Left hand pulling the left throttle down, killing fuel to the engine on fire, seeing through to the new problem he's just created. Killing power costs him airspeed—the enemy can recognize the change as sign of a vulnerable prey and come back to finish him off—so he pushes the right throttle all the way forward, easing the one working engine to full military power, just short of afterburners. Burning through more fuel than he can afford to, but that's a problem for a few seconds from now.

A city appears through a break in the cloud cover.

"That," Hilton says over the intercom, "is Hanoi." He points down. Hanoi is the stuff of nightmares. If they eject now, they both know what'll happen. Assuming they survive the ejection, they'll parachute right down to the angry men with machine guns, who will break their arms and torture them and take them to a prison to torture them more. "I'm not going."

He figures they're perhaps seconds before the rear of the plane explodes. They might be able to survive that. He thinks through the mechanics.

"When the tail burns off," he says, "the bird will pitch up." The center of gravity will shift and he'll lose control. "*That* will be the time for us to go. The pitch could be violent, so be ready."

He keeps the right throttle forward. The fuel gauge winds down.

Around him, seven planes are now spiraling out of the sky, driving his friends into the ground. This doomed mission, this harebrained idea to bait MiGs into the air because he's not allowed to bomb them when they're parked on the ground and harmless, not allowed to bomb the SAMs before the SAMs are wired up in their hidden nests and at their most dangerous. The politicians won't let him; the politicians would rather risk Hilton's life, Mike's life—all of their lives—because the politicians want to win a war without really fighting a war, and there isn't a weapon, so far as Hilton knows, that can do that.

HE SCANS THE horizon. He can see the Red River, and a dense swath of forest just beyond. It'll do. If he can stay aloft for just a few more minutes before they have to eject, he and Mike will have some cover down there and the chance to hide for a while. Hilton lets in a glint of hope.

He hears a voice in his helmet—a pilot flying alongside them: "Fire's blown out."

That's good news. Hilton thinks it probably went out because it couldn't get enough oxygen this high.

Or, he thinks, because it ran out of fuel.

He checks the gauge: he's lost 90 percent of his fuel supply. A minor crisis that yields immediately to an idea: Why not try to get some more?

He clicks at Mike over the intercom. "Let's go Brown Anchor frequency—"

Before he's finished speaking, a voice comes back but it's not Mike; it's someone else, giddy and eager, like its been waiting for this moment.

"Don't do that! I'm heading north and got her buck'n'!"

And now it's hard not to smile: it's the tanker crew, already coming to the rescue. They're supposed to be nowhere close to here, orbiting in safe airspace, but they must've listened in on the massacre. They've ignored a standing Strategic Air Command order to stay away from high-risk airspace, and now they're risking court-martial and maybe their lives because there's one plane that might just have a chance.

"When do you want me to turn?" the tanker asks.

Hilton runs numbers in his head. To refuel, they have to be flying in the same direction, and right now they're likely flying more or less right at each other. For a head-on approach, the tanker should turn at twenty-six miles, he knows this from the manual, but this tanker's likely flying way faster than normal, and Hilton is running out of time.

He radios back: "Turn *now.*"

He glances at the fuel gauge again. A thousand pounds . . . nine hundred. Maybe two minutes of flying time.

Hilton thinks ahead. He figures he'll have only one chance to hook up with the tanker. If he's too slow, he'll run out of fuel before they connect. If he's too fast and makes a mistake, he won't have fuel for a second try. He sees a flash off the horizon, a rounded surface. The big mid-turn tanker showing its belly to the sun, and now he knows there's a chance.

"Slow it down to three hundred knots," he radios.

"Coming back now."

Hilton feathers the stick, subtle adjustments. Seven hundred pounds of fuel left . . . five hundred, basically vapor. He's closing in now, the tanker growing in his canopy glass. Four hundred pounds of fuel . . . three fifty; he braces for flameout. The refueling boom emerges from the back of the tanker, groping for the small receptacle next to the canopy a few meters behind him. Two hundred pounds, and his one chance.

He nudges the Phantom forward. The refueling boom bangs against the fuselage and slips into the receptacle, and as the gauge drops below two hundred pounds—nothing; a rounding error—the Phantom begins taking on fuel.

Hilton allows himself one more moment of release; he has a moment for reflection, to contemplate why he's alive and turn to the needle of irritation already beginning to fester into something harder. He doesn't yet know how many of his friends are dead, for no good reason. Because of a foolish plan they'd been duty-bound to follow.

The gauge passes three hundred and fifty pounds, the plane's forward tanks almost full.

He doesn't know why he and Mike are still alive. He still thinks luck or providence—something someone else controls, nothing special about him, surely, though he knows he's maybe a bit of an aberration as a pilot. Not everyone can sit in a cockpit and become *part* of it, a brain encased in it, translating speed and violence into data points to examine and rearrange. The gauge passes five hundred pounds of fuel.

So maybe it was some hidden instinct of his own, invisible even to him, that steered them through the SAM storm, and he's starting to wonder if maybe that's the reason others are gone now, down there, spread out around their broken planes, while he and Mike are still alive.

The gauge passes 1,000 pounds, a violent convulsion rocks through the cockpit, and Hilton hears himself yelling into the radio.

"Breakaway. Breakaway."

He yanks the throttle back, slamming the brakes so the boom flies forward into empty air, the tanker pulling away and sputtering jet fuel into the sky. He waits for someone to yell "Fire!" and keeps his distance but knows he can't wait long. His fuel is already dropping below four hundred pounds again, and now the problem is compounded.

He needs more fuel or they'll fall out of the sky; if he takes on more fuel, he risks an explosion. He pictures his plane's ruptured innards; he scrolls through his memory, trying to recall diagrams from the flight manual, black-and-white sketches from classroom study days. He sees the stenciled arrangement of fuel tanks, six of them lined up front to back inside the fuselage—and it clicks. The SAM detonated near the rear of the plane. The problem is with the rear tanks. They must have been punctured by shrapnel blasting through the fuselage. The forward tanks must be intact, but when they're topped off, fuel begins flowing into the rear tanks, and the moment fuel hits the rear tanks, it's being shot out into the slipstream.

Hilton radios the tanker crew. "I've got an idea."

He eases the throttles forward again. He nudges closer to the tanker, centering on the refueling boom. He manages another near-empty, near-perfect rendezvous. He starts taking on fuel, waits a moment, and radios back to the crew.

"OK, cut it!"

He's taken less than eight hundred pounds, only enough for a few minutes of flight time, but at eight hundred pounds the forward tanks aren't full yet, so nothing is flowing into the ruptured rear ones. Still connected to the tanker, he burns almost all the fuel down until he's nearly empty again.

"Back on."

That's how they're going to get back to base: they'll stay connected all the way over North Vietnam, all the way to the base in Thailand, a kind of tandem flying none of them have ever tried before, spigot on, spigot off, for a few hundred miles, and soon the base is almost in view and it seems, for the moment, that they're finally home free.

"OK," Hilton radios the tower. "We're coming in. Clear us."

He checks his instruments. His airspeed is too high, but he's lost the hydraulic system that deploys flaps—and that extends the landing gear and controls the wheel brakes once he lands. There's a work-around he's never tried, but it's his only option. He pulls a handle that levers a bottle from the side of the plane out into the slipstream, to accumulate air pressure like a balloon filling up. The odds aren't great that a balloon blowing air through the plane will actually work in place of hydraulic fluid, but it's all he's got. He grabs the flap handle and pulls. The flaps lift up from the wings.

He reaches for the landing gear handle and pulls hard. He can feel a clunk and a mechanical grinding—doors below him levering open, up locks releasing—and there's change in how the airframe is moving through the air. The landing gear is down.

All there's left to do now is pull the nose up and flare to landing, but it's at that moment that another flight control system Hilton didn't even know was damaged leaks the last of its fluid, and Hilton loses control.

The plane drops the last ten feet with no one controlling it. The wheels slam into the ground, the landing gear survives impact, but the plane is now racing across the tarmac, a fifteen-ton tricycle chewing up runway far too fast. Hilton yells to Mike to start unstrapping, he's going to open the canopies even as the plane is still moving over a hundred miles per hour, he's going to jam down hard on the brakes and hope they work, but if they work, the last of the fuel will lurch forward toward hot parts of the plane, and everything will ignite. They need to be ready to jump. He begins thinking through the maximum speed at which a human body can survive the leap from a moving plane. He gives Mike one more warning, then jams his feet on the pedals as hard as he can. The brakes grip; the tires skid. He braces for ignition, the wheels gripping, ground speed leaking but maybe not fast enough. He squirms out of his seat, and the second the plane shudders to a rest, he's out of the cockpit. He meets Mike out on the wing. Hilton feels himself finally downcycling from machine to human being again. Now he feels the adrenaline; suddenly he's nearly giddy with relief.

"Mike!" he says. "If *I* had ninety-nine missions to go, I think I would cut my throat."

Mike doesn't laugh. He's wide-eyed, exhausted, hyper, in some kind of altered state. All he can muster is "Do you think God was with us today?"

5.

Ubon Royal Thai Air Force Base
U.S. Air force 433rd Tactical Fighter Squadron station

After he showers and cools off, Hilton gets angry.

Friends of his are dead. The mission accomplished nothing. The MiGs are still out there, untouched; the SAM sites are still out there, untouched. The supply line is still running, the Dragon's Draw bridge is still standing; he's done nothing to interrupt the supply line feeding the Viet Cong fighters killing Americans.

Back on earth, Hilton stages small acts of resistance mostly for himself. He avoids the officers' club; there's no love lost with the top brass, so some of the guys in the 433rd scraped together their own improvised bar where they didn't need Permission to Speak Freely: first a couple of planks in some spare space between two bunks, then decorations just kind of drifted in. The 433rd was nicknamed Satan's Angels, so they called their little hooch the *Inferno.* And it was mostly there—drinks fifteen cents, a quarter for top-shelf, all of it on the honor system except for when a local guy named Ot showed up to tend bar—that the Satan's Angels moaned. *The politicians want it both ways,* stumbling pilots shouted in each other's ears. *To win a war without really being in a war.* There was no way to do that yet, so young pilots went up in the sky and died, and on humid nights—through the fug of heat rising off the tarmac, weighted down by cold beer going warm—drunk pilots who'd stumbled home in shot-up planes sang laments and spilled liquor for their fallen brothers.

When he wasn't flying or in some shoulder-locked prayer with swaying pilots at the *Inferno*, Hilton tried other ways of distracting himself.

He had Ann back home to think about, maybe worrying, maybe knowing by now not to worry so much: from the letters Hilton forced himself to write she had enough data to know there were ways the place tried to redeem itself. If you could put the frustration aside, and the fear, you could fool yourself into thinking life here wasn't all bad. He liked what there was to drink, he liked the food, the heavy French sauces at the on-base Thai restaurant called Thai Restaurant. He liked "wolleyball," which they'd proudly invented even though it was just volleyball plus cheating. He didn't like writing letters, but wanted Ann to hear from him. Someone once suggested a tape recorder, so he started sending tapes home, and Ann started sending tapes back. When he found even speaking aloud couldn't quite capture what it was like to fly these kinds of missions, he wandered into the machine shop and asked the mechanics to rig the recorder into the cockpit. Soon Hilton was hooking Ann right into his comms, taking her aboard. He'd get in trouble if the brass found out, but he had the old soldier's excuse: *What are they gonna do, send me to war?* For Hilton, and the rest of the pilots at the *Inferno*, little acts of mischief served as proof they had some control, ways of maybe worrying a little less about being entirely subject to the whims of politicians, worrying a little less about missiles and MiGs and supply lines, worrying a little less about the Dragon's Jaw.

Hilton listened to the stories of the Dragon's Jaw and waited for his turn to be sent against it. It always circled back to that damned bridge, whatever it was—hologram, anomaly, or just a monster, protected by monstrous things. The pilots who survived it talked about going in low, pushing in close to the ground. You had to get right down on top of the bridge to have any hope of hitting it; you had to push down through the field of antiaircraft fire coming from all directions. The North Vietnamese had every weapon in their arsenal defending that bridge: MiGs and SAMs, but also antiaircraft artillery in all different flavors—radar guided, optically sighted, guns nimble enough for two people to operate, guns so big they required a team of eight or more running along the river with ammo boxes. There were guns installed on the riverbank, in the villages nearby; guns on the hilltops overlooking the bridge; gun positions mounted on a barge floating in a rice paddy—all set up in geometric patterns to create overlapping fields of fire. Flying against the Dragon's Jaw meant flying through grids of explosive metal, diving through all of it before releasing your bombs to have even a chance of hitting the bridge.

And if your plane was somehow still intact after dropping your bombs, you had to pull up hard to avoid the ground, but not so hard that multiplied force of gravity stripped the wings off the plane. Even for pilots who pulled all that off, average accuracy was laughable. If you hit the pickle button a half second too late, your bombs missed the bridge by at least a hundred feet. So pilots rarely got their bombs within a hundred yards.

Thousands of pounds of bombs were spent trying to take that one bridge down. Seven hundred sorties, twenty-nine planes lost. Pilots lost their lives, or their planes, or were shot down and dragged to prisons in Hanoi to be tortured. For the airmen at Ubon, covering their eyes and looking at the specks on the horizon to identify the planes that survived this time, the whole war was that bridge. The whole world was that bridge. It resisted everything. Bombs fell, the smoke cleared, the bridge stood, a mythology rose up and hardened. More than 90 percent of the bombs sent against the Dragon's Jaw missed.

Mission planners began dreaming up wild ideas: sending dozens of planes at a time, crowded-mall parking lots in the sky, each pilot waiting in dangerous airspace for his turn to dive. They tried small strikes with fewer planes and less waiting around, but those strikes did almost no damage. They tried every tool anyone had ever come up with to try and get a bomb to steer on its own, a concept the military had been interested at least since the time in World War II when they attached a television camera to the front of a B-17, loaded the plane with explosives, and tried to fly it by remote control at a Nazi vengeance weapon bunker. The latest version of Joe Kennedy's steerable bomb was the "Bullpup," a missile with a television camera. A plane had to follow close behind to keep the Bullpup flying toward the target, so it was still dangerous, and anyway, the bomb's mass was made up mostly of rocket motor, a big weapon with a small explosive package. Bullpups failed to drop the Dragon's Jaw. The United States had steerable surface-fired missiles like the SAMs the North Vietnamese had, but they weren't very precise and were really only effective against fragile targets like planes, since, unlike a hard target on the ground, a plane could be destroyed without a direct hit, even by shrapnel from a nearby explosion.

They tried electro-optical bombs, which could theoretically "see" targets using the contrast between light and dark. A sophisticated idea that didn't work very well in practice. Electro-optical bombs lost their targets constantly, and even in the best case could only operate during certain times of day when the shadows were just right.

They tried hitting the Dragon's Jaw from below: the Air Force launched a top secret project to design mines that could float down the river and explode underneath the bridge.

A cargo plane took off with its rear doors open under cover of night, flew a few miles upstream, and let the eight-foot mines slip down rollers and parachute into the river. The crew returned to base proclaiming the mission a success, but a recon flight the next morning showed no damage to the bridge. Another mission was ordered for that evening, and this time the result was even more mysterious. The plane never returned to base. The crew was never heard from again but was eventually assumed to have been shot down near the bridge. It was unclear what happened to the floating mines, but again the bridge was still there in the morning.

They tried "dual axis" attacks to try to split the bridge's defenses by approaching from two angles; they tried joint Army-Navy operations, elaborate, multistage missions to take enemy radars offline before attacking. Bombers bombed for four hours, wave after wave of bombs coming in sheets like sleet storms. Down on the ground, girders twisted, railcars crumpled, but every time, when the smoke cleared, the bridge still stood.

Hilton tried to explain all of it in the tapes to Ann. What it was like there, biding his time before flying against an impossible target. He resolved to record as many tapes as he could. Talking about this mythical bridge, and talking about inane things too: the food, his friends, the rib he might have broken with the Thai boxing teacher—a library of anecdotes so she might understand what it was like here and to build up an archive of his voice saying nice things to her. Because soon it would be his turn to fly against the Dragon's Jaw. And when he did, Ann might not get any more tapes.

BOOK II

WELDON

(OPERATION LINEBACKER)

1.

Detachment 5
Eglin Air Force Base

After every mission against the Dragon's Jaw, just after the last surviving bomber pulled out of its dive—after the guns on the ground had gone quiet; after the radars on vans at SAM sites stopped spinning and the MiGs returned to base—one last American plane angled in. This plane a little different from the others. A longer nose, a little faster. It passed high and safe miles over the bridge, with a rig of high-powered cameras to capture the landscape after the attack, then curved out toward the sea and returned to base.

Down on the flightline, a crew member opened a compartment in front of the spy plane's cockpit, removed the film cartridges, carried them inside to be developed, and one photograph so starkly showed the failure at the Dragon's Jaw—a cratered moonscape, everything stricken but the bridge—that it was slipped into a special pouch, marked CONFIDENTIAL, and shipped to the Pentagon. At the Pentagon, the pouch was unsealed, the photographs reviewed and forwarded to U.S. Air Force Systems Command, which was tasked with overseeing research into new weapons. Systems Command forwarded it to the Aeronautical Systems Division, which forwarded it again to the new Limited War Department, where an officer decided to forward it a final time to a brand-new, small, secret group at Eglin Air Force Base in Florida, a group known to the few who were aware of its existence as Detachment 5.

There, the photographs found their way onto the desk of an overenergized Air Force colonel who'd just been put in charge of the group, who was already apoplectic about the military's failure in Vietnam, and who was beginning to feel like he might be willing to try just about anything.

WELDON WORD WALKED out of the colonel's office with a sense that the colonel was willing to give him a chance. The colonel seemed intrigued by the idea of a weapon *guided* by laser and seemed unbothered by the fact that that it sounded outlandish. Maybe it was appealing *because* it was outlandish. None of the conventional ideas had worked, after all. Nor did the colonel seem concerned about the fact that Weldon's company, Texas Instruments, had basically no experience in making weapons.

But Weldon soon found out the colonel was perhaps the only person on base who felt that way. Nobody else seemed to think that talking to Weldon was worth their time. In the days after the meeting, as Weldon went around the base looking for advice, engineers from other companies shooed him away. Air Force officers pretended to be busy. The moment he began talking about lasers, half the engineers laughed him out of the room, and the other half asked what lasers were. Few people on base felt there was anything wrong with the kind of weapons they were already developing, and anyway, if anyone was going to build a new weapon to solve a problem like the Dragon's Jaw, it was going to be one of the big defense companies that already had the answers to the questions Weldon was asking. North American Aviation could put hundreds of thousands of dollars into developing a weapon. They'd built fighter planes; they built jets that could do Mach 3; they'd made the P-51 Mustang fighter, which had arguably won World War II, and the B-25 bomber, which had also arguably won World War II. Texas Instruments had built a chip. They were electronics nerds.

After three months trying and failing to convince other Air Force staff to take him seriously and help him develop a prototype, Weldon called his bosses in Texas and told them it was probably a waste of time. Then he went to see the colonel so he could graciously bow out. He told the colonel it was time for him to move on. It wasn't the right environment for a small outfit like Texas Instruments, and everyone seemed to like things the way they were. "No one else at Eglin feels the Air Force has a bombing problem," he said.

The colonel hit the roof: this was exactly the kind of conventional stuck-in-their-ways thinking that was getting pilots killed, all the rigid thinking that was losing the war, and when he calmed down enough to string an intelligible sentence together, he made Weldon a promise.

"Have a proposal on my desk by 7 a.m. Monday and it's as good as funded."

Weldon had no time to ask how keeping a promise like that would even be possible because the colonel was already running through a list of impossible specifications: the new weapon should be ready to test in under a year—make that half a year. It had to be able to reliably strike within thirty feet of a target, and—given the restrictions on Detachment 5—Weldon had to pull all that off on budget of under $100,000.

Weldon had less than seventy-two hours to come up with a plan.

2.

At thirty-four, Weldon was already a father of four children who no longer had a mother. Four kids who were far too young to be motherless; their mother had been far too young to die. It would take him years to figure out how you're supposed to talk about a thing like that. When another engineer brought up his own family and Weldon tried to relate, he stalled: "Look, let me tell you, this stuff is important. You don't want to have . . . you don't want to . . ." He wanted to be helpful and didn't actually have anything usable to say.

But he'd found you don't always have to say something usable. That he sometimes got further the less talking he did. He'd been given a physical gift he was just beginning to learn how to use: a prized set of big, bright blue eyes, the brightest in the room. When he turned to colleagues, when he looked right at them, they saw a man listening so intently, they'd sometimes wonder whether anyone else had ever really listened to them at all. He was learning that often all he had to do was settle his eyes on people; often he didn't have to say much at all.

When he did speak, he had a deep, imposing voice, and all of it—his eyes, the silence, that reverberating voice—could make people think their questions were being considered by a council on Mount Olympus. He was the kind of person outwardly so composed, he appeared to simply be above the frustrations of mortal life, unruffled and ready to dispense wisdom on a low-temperature setting as though he'd survived the particular crisis you were talking about and made it through without breaking a sweat. Grace and composure, and I'll tell you how it turns out if you give me a minute to think.

It wasn't entirely true, of course. He wasn't totally in control—not of everything. Not of some of the most important things. There were things beyond the reach of the most potent, focused mind. Elements of physics and history, rules of science that broke for no one. He had no control over biology, over improperly splitting cells. The chance transcription error that repeated itself and moved through his young wife's body, variables you couldn't solve for. Huge ones, cascading ones, that could steer a person toward trying to control the things he thought he could.

So Weldon found himself drawn toward problems he thought he could solve. He'd attack this one. He'd bury himself in it; he'd use his shiny eyes and the whole spectacle of his listening to draw people to his cause, and he'd find a way to build the colonel his impossible weapon.

He found a math whiz stationed at the base who could afford to lose a weekend, too, and convinced him to take on the challenge, and together the two of them stayed up for two days straight, scrabbling together a rough idea for a bomb Weldon knew would never fly. He wagered that it might not matter: he saw in the colonel a highly motivated man who just needed something to work with, so Weldon and his new recruit spent the weekend sketching designs in pencil, then redoing the least ridiculous one in India ink, working the problem over and over until an odd-looking machine finally began to take shape on paper. A clumsy contraption, inelegant and unwieldy even in two dimensions, that looked less like a projectile than a piece of scaffolding. Struts stuck out off a bomb, holding a small cylinder that would be, in effect, the bomb's "eye."

As they refined the design, the eye emerged as the key idea. After all the fuss in the Army about lasers, all the tests and demonstrations, there was still one use that hadn't yet been fully explored. Weldon's idea for a precise, long-range weapon for the Dragon's Jaw took the Army's concept of a not quite death ray and tweaked it, dispensed with the idea of trying to make laser better at what it was bad at—damaging things—and embraced the less dramatic thing it was already good at. Weldon's idea didn't require a laser powerful enough to destroy the bridge. He just needed a laser powerful enough to point at it. Somewhere, someone would hold the laser and shine it, almost like a flashlight, at the bridge. On the bomb, the cylinder—the eye—would look for the laser reflection, somehow be able to see it, and somehow, using TI's chips, would know which fins to kick to steer toward the target.

After a seventy-two-hour dash through the weekend, Weldon brought eighteen handwritten pages of a not-yet-fully-fleshed-out concept to the colonel's office. The colonel leafed through the proposal.

"I'll take it from here," he said. "Now get lost."

Weldon bought a six-pack and headed to the beach.

A FEW DAYS later, the colonel called Weldon back to the Detachment 5 office. Good news and bad news, he said. He'd sent Weldon's proposal up to Air Force HQ in Ohio, where the Project 1559 folks saw it as just the kind of wild idea the Air Force was trying to entertain. They'd given the green light for further development.

The bad news was that the Air Force was hedging. They were giving him only two months to see if he could make it work, and since there was little reason to believe an upstart like TI could really pull this off, Air Force brass was also sharing Weldon's proposal with North American Aviation so they could develop a prototype as well. If anyone was going to make the outlandish idea work, it wasn't likely to be a young engineer from a chip company whose closest contact with military contracting thus far was collecting sensor data for an anti-submarine sonar program off the coast of Rhode Island.

Weldon got to work. He knew that if Texas Instruments was going to beat North American Aviation to the finish line with a working prototype of a laser-guided bomb, he was going to need help. He needed to expand the team. He called Dick Johnson, a former test pilot with a hobby of flying gliders and a near obsession with data, and Jack Sickle, a retired Navy pilot who had so much pent-up frustration with overcomplicated, finicky weapon systems that he'd come to see simplicity as something like a divine ideal. And he knew what it was like for Air Force mechanics on flight lines in theater. "Keep it simple, because you're depending on kids two years out of high school," he said to Weldon once, then repeated, again and again, "to put this stuff together and make it work."

By inviting Jack Sickle to help, Weldon was making his own job twice as hard. The challenge was no longer building something complicated; it was building something simple. Sickle's refrain shifted Weldon's focus away from pleasing generals and more toward the last stop, where kids on ground crews decided whether they liked a weapon or not. If they liked to use it, Sickle insisted, they'd find a way to use it. If it was confusing or frustrating, they'd find a way not to. The weapon had to be *easy*. Easy to assemble, easy to fix, easy to use.

There were a million problems to solve. How to build a brand-new weapon compatible with old planes? How to get fins to steer a bomb? Bottles of compressed air could move fins, or gas generators could, too, but how do you attach them to a bomb? All while Jack Sickle was picturing himself on a ground crew, imagining himself in a cockpit, kneecapping each idea Weldon and Dick Johnson offered to solve each problem. "Too complex—it's too complex," Sickle said, nearly monastic in the consistency of his refrain. "Make it simpler."

Most importantly, how would the "brain" work? An unpiloted projectile capable of steering had to be, in some primitive way, an unpiloted projectile capable of "thinking." This wasn't the first attempt. The Anvil project had long been abandoned, canceled after the mission that killed Joe Kennedy Jr., and even if it had worked, it would have required destroying a plane with every mission. Steerable projectiles effective for other applications thus far weren't helpful for this one: the Redstone rockets that Tom Weaver's team of ex-Nazis worked on were expensive and complex and required huge teams of people and buildings full of equipment for each launch. Surface-to-air missiles like the ones terrorizing American pilots over Vietnam could steer without anyone aboard but relied on dozens of networked vehicles and people and a handful of radar systems on the ground—all told, a few million dollars of hardware—to tell the SAMs where to go.

To give the colonel what he was asking for, a weapon easy enough to use that pilots could operate it more or less by themselves, Weldon needed to come up with a weapon that could "think" more or less for itself. Even the self-steering weapons had been tried before—even the ones that worked to some degree, like the early air-to-air missiles, or the electro-optical and anti-radar bombs—were all too complicated, or too unreliable, or just too dependent on specific conditions to be useful. There still wasn't an autonomous weapon that could "think" reliably or independently enough for the colonel's purposes, but that was one problem at least that fit squarely in TI's wheelhouse. Getting into the weapon business was, after all, partly just an excuse to create a robust market for TI's new "chip" made from silicon. Weldon already had something that could make simple decisions—he had the brain—but he didn't have the nervous system or the body.

If the weapon was going to steer toward a target, it would need to "see" the target designated by the laser, which meant it was going to need an "eye." What material would the eye be made of? How would it work? How do you turn sight into thought and thought into movement? It took weeks before they stumbled on a solution that passed Sickle's s test: they could make an eye out of a photon detector, a lens that released electrons when exposed to small particles of light. When light from the laser reflection—however weak, however many miles away, even at a frequency the human eye couldn't see—hit the photon sensor, those released electrons would produce a signal. They could wire the photon sensor to their new Texas Instruments chips, which could block or pass on the signal, amplified so it had enough power to trigger a gas canister, which in turn could release a blast of pressurized gas strong enough to move a fin.

The next task was figuring out how the bomb would know which way to steer, but that turned out to be a deceptively simple problem to solve. It was all in the wiring. Each bomb would have, in effect, different parts making up its eye: the photon sensor lenses to the left would be wired to chips that were wired to gas canisters that were, in turn, connected to the fin that, when triggered, would steer the bomb back toward the left. Just as a person who sees an object moving toward the left of her vision knows she's moving to the right of it, the bomb's eye would "see" the laser reflection moving to the left, which meant the bomb was off course to the right. Each part of the eye would be wired to steer the weapon back in its own direction. The effect would be of a bomb thinking for itself, though, really, Weldon's team was doing the thinking for it and leaving only four options available to the bomb: steer up or down, left or right.

This concept, too, proved sufficient for the Sickle simplicity test: cumbersome though the wiring might be, if it worked, it would work simply. The bomb would fall through the air, constantly correcting, trying to get the reflection toward the middle of its vision. And, once dropped, it would do the work itself, without the pilot having to steer it.

THE CONCEPT SEEMED workable, but there was still one major problem, one that had troubled attempts at "steerable" bombs since the beginning. Even if you made an eye that could see a target, it would only work if the eye was facing the target. Bombs falling from the sky didn't stay in consistent positions or fall in perfectly straight lines. They wobbled in the air, blew off course; they'd twist and buck around their axes. The moment a bomb was yanked off target, its eye would be yanked off target, too, and bombs couldn't steer toward laser reflections they could no longer see. Weldon and his team needed a way to *keep* the eye looking toward the target even as the bomb rolled and yawed through the air.

Perhaps they were being too faithful to nature, deriving the idea of an eye too rigidly from what they saw in living creatures. If you stopped thinking about eyes anatomically, it freed you to receive new inspiration.

What if the eye was way bigger than you might find on an animal? What if the eye wasn't removed from the body at all but actually *was* the body? If you could turn the whole surface of the bomb into one big, panoramic photon-sensing lens, then the target would always be in sight no matter how the bomb rolled through the air.

Almost as soon as they broached the idea, it revealed itself as impractical. Even if they could solve the engineering problem, there was no way to solve it in under two months, and even if they did, the materials for a photon-sensing lens that large would be prohibitively expensive.

If they couldn't build an eye big enough to maintain an unblocked view of the target throughout flight, the only other solution was to somehow give the eye the ability to look around, which meant giving it a head and a neck, too—some kind of robot component. That would perhaps be doable with a budget like one of the real defense contractors had—North American Aviation was probably building a robot head for their version—but Weldon's team didn't have anywhere near the resources North American did. And he had the added challenge of Jack Sickle, one foot back in his fighting days, picturing a kid sitting under a fighter jet, blinking dumbly at some elaborate contraption. "Keep it simple. Gotta keep it simple. Keep it easy to fix."

They were building a bomb to steer itself toward a laser reflection without a way of looking for the laser reflection. With time running out and North American Aviation surely well ahead, Weldon needed a stroke of genius, and he needed it quickly.

3.

The team was gathered on a Saturday afternoon in a lab near the Detachment 5 office, trying to divine a solution to the robot head problem, working through the weekend because there was no time to spare, and anyway, weekends were the only time they could get lab time. As they crunched numbers with growing urgency, Weldon could see Dick Johnson's mind beginning to wander, the former test pilot sifting in his head through the whole history of flight innovations he'd surely encountered over the years. No one on Weldon's team personified the Texas Instruments self-image more fully than Dick Johnson: half maverick, half nerd, an adventurer with a slide rule. Dick had come up as a teenager with a square haircut and spectacles, flying gliders he assembled himself. He'd learned to love data as much as adrenaline and saw the tools used to test inventions as worthy inventions themselves, some of them almost as impressive as the very act of flight.

The pitot tube, a device with a hole on the front and one on the side and mounted to the side of a plane because someone figured out you could use the difference in pressure to determine airspeed. Instruments that used air venting in or out of a diaphragm connected mechanically to a dial to show how quickly a plane was gaining or losing altitude. The Giannini probe, sticking out from the side of a plane, attached but moving freely on a gimbal and shaped like, of all things, a badminton birdie so it stabilized itself just by the flow of air. The Giannini probe would always point in the direction of motion, so it could help measure the difference between where a plane was going and where it was *pointing.* Brilliant devices collecting numbers and figures Johnson could pore over later. Testing planes was a data lover's dream, a place with all the adventure of flying, but also hidden stories that emerged if you looked closely enough.

After World War II, Dick enrolled in a college engineering course and took on as a mentor an aerodynamicist who reinforced a lesson Johnson was already starting to intuit: that often the best inventions are only half-invented. Often they were just new uses of things originally built for a different purpose. Johnson began applying the mindset to the gliders he still built and flew in competition. He looked around for ideas to borrow. He quickly found a big one: a new wing shape used to smooth out turbulence on fighter planes, which was about as far from a glider as you could get and still be flying, but Dick figured drag was drag whether you were in a powered aircraft buzzing over enemy territory or a silent glider floating above some quiet southwestern desert. So he studied the idea, modified his glider with a "laminar flow airfoil" he saw on the fighter planes, won a distance competition, and beat the state of the art by about 20 percent.

He hadn't even needed to invent a wing design, just go looking for it. So, a few decades later, when he took on a project with Texas Instruments working for a recently arrived engineer named Weldon Word, who settled his bright blue eyes on Dick and asked for help solving an unsolvable problem with almost no money and less time, Dick had an idea of how to approach it.

In the lab, 3:00 p.m. came and went, then 3:30, then 4:00. Weldon had his family to check in with. Sickle had whatever Sickle had planned for Saturday night. Dick had a pool to get home to.

Five p.m.

Dick was thinking, *The best inventions often aren't inventions at all.* In Dick's head, the bomb transformed into one of the aircraft he used to test, into one of the gliders he used to fly in competition. What was a bomb, after all, but a glider that didn't glide very well? An unpowered projectile moving through air?

Six p.m. passed. Then 6:15.

They needed some kind of *head.* A shape connected to another shape. What inventions already existed for gliders? A shape that could move freely but stay pointed in the one direction that mattered?

At around 6:30 the idea leapt from Dick's mouth at almost the moment it sprang into his head.

"The Giannini probe!" The badminton birdie–like device that stuck out from the side of a plane. "Let's put this seeker in a Giannini probe!"

It was suddenly obvious. Air moving through a badminton birdie aligned it in the direction of motion; birdies wobbled for a moment but always settled into a steady alignment. That was why when a plane flared to land, descending but with its nose pointed up, the Giannini probe still pointed down, in the direction of motion. The team didn't need to build some kind of complicated neck or robot head to keep the bomb's eye on the target or figure out how the eye would know where to turn. They didn't need some contraption built from gyroscopes and servos and motors that on its own would cost hundreds of thousands of dollars. They needed a $300 universal joint to serve as the neck and a head shaped like a birdie. Even if the bomb wobbled off its axis, its head would still be stabilized in the air, turned toward the target. They didn't need to figure out how to turn the eye toward the target; air moving across the falling bomb would do it for them.

4.

Project 1559 headquarters
Wright-Patterson Air Force Base
Just outside Dayton, Ohio

Weldon sat in a conference room in Ohio, finishing a presentation the colonel dispatched him to give, pitching the Project 1559 higher-ups on the masterstroke that might just revolutionize warfare. He wrapped up his spiel, laid out the diagrams, and awaited questions.

The officers looked at each other. The room erupted in laughter.

He should've expected it. The idea of lasers was already hard enough to take seriously, and the whole design in the sketches he presented looked more like some kind of lawn toy than a weapon. Dick's ingenious Giannini probe breakthrough didn't look very elegant on paper: a birdie at the end of a long strut sticking out from the bomb like a child flying a toy out a car window. The Air Force brass was unmoved, and Weldon's two months were up.

But Weldon was a horse that the hard-charging colonel had decided to back, and if the brass wasn't exactly impressed by Weldon's toy shop design, the whole purpose behind the Air Force standing up Detachment 5 and Project 1559 in the first place was still to try a novel approach to defense contracting, seeding even bold new ideas with small investments. The brass conferred and eventually agreed to give Weldon $99,000—less than the unit cost of most weapons—for the entire project, including twelve working prototypes.

They'd technically been given a chance to succeed, but it almost felt like they were being set up to fail, and there was one more caveat in the funding agreement that perhaps explained why: If Weldon failed to produce the working prototypes, the Air Force would order TI to hand all their progress to North American Aviation or one of the other *real* defense companies.

The colonel seemed to believe the little upstart from Texas might be the team to actually solve the problem of the Dragon's Jaw. Almost no one else did.

BACK IN FLORIDA, Weldon's team started attacking problems. They needed to find clever ways of saving money. It would have to be bootleg.

They figured they could move faster if they skipped building the part that actually exploded; they left that to other companies. Their bomb wouldn't really be a bomb but a kit that attached to someone else's bomb.

On their minuscule budget, they decided the fastest way to build a working prototype was to channel Dick Johnson's approach: invent by inventing as little as possible. They cannibalized parts of other weapons they found around the test range. They borrowed electronics off a different missile; they took the fins off other bombs. A Frankenstein's monster of scrapped parts began to take shape, but they ran into a new problem: how to test the weapon. They needed wind tunnel time, but that was getting harder and harder to come by. Around them the Space Race was heating up, and bigger companies with bigger budgets were working on major projects, faster planes, and ballistic missiles. It felt like they all needed wind tunnel time, and they all had bigger budgets than Weldon's team did.

With a rough design more or less complete but progress now stalled, the team disbanded one evening and Dick went home to sit by his pool and think.

And out there, it occurred to him: *the pool.*

Dick turned around and went back to work. He machined ten-inch models of the bomb and hooked ropes through them. Back at home, he dropped the models in the pool and began dragging them from one end to the other, watching how they moved through the water. He reported to Weldon: he couldn't be 100 percent certain without a wind tunnel, but hydrodynamics were a decent approximation of aerodynamics, and—based on the way the models moved through the water—it looked like the design was sound. Sideways pressure was only a few miles per hour, and performance in the air shouldn't be much different. The bomb should be stable in flight.

This wasn't just engineering; this was jazz—constant improvisation over a chorus of rules they'd learned and then learned to break. Each successful improvisation a jolt of confidence that they could find a workaround for just about anything.

Next they had to test the laser. It was hard to know if the laser was working because it was invisible to the human eye, and the bomb wouldn't work if it didn't have a laser reflection to look for.

But a laser was just light, and even light humans can't see could expose film. At the time, Polaroid was in the middle of a push to bring its newest "picture-in-a-minute" camera to the masses, making the cameras and special film available for sale in drugstores across the country, airing television ads with movie stars and a Barry Manilow jingle, and partnering with famous artists like Ansel Adams. Anyone paying even passing attention had seen Polaroid's rapid self-developing photographs in action.

Rather than inventing a sophisticated sensor to test their laser designator, the team tried taping an undeveloped square of Polaroid film up on one side of a room and aimed the laser they'd use from the other. They turned it on, then off, picked up the Polaroid, peeled off the plastic backing sheet, and, sure enough, a ghostly white smear emerged. The laser designator had exposed the film. The laser was working.

Soon all the major parts of the bomb had passed some jury-rigged test or another, each on its own. Now it was time to put the whole thing together.

ON A MIDSUMMER day, the band of misfit engineers gathered at a telemetry pad before a massive test range littered with decades' worth of spent ordnance, looked up at the sky, and prayed as a test pilot called out his final approach over the radio.

5.

Detachment 5
Eglin Air Force Base

A million things could go wrong now. The bomb might fail to detach from the plane. It might not stabilize in the air. The fins might not move. They had the laser mounted on a tripod and pointing at the target; the laser could fail. Weldon's team hadn't built that part, but if the laser stopped working, the bomb would miss the target and they'd be blamed.

They'd had to forgo the onboard radio equipment that big defense companies used to collect data in real time—radio equipment was beyond their budget. Instead, they packed sensors into a coffee can–sized container and stuffed it into the inert bomb. The sensors might not survive the impact, or they might not even find it.

The pilot came over the radio to signal he was releasing the bomb.

Up in the sky, a speck separated from the plane—weapons release successful—and crawled toward the earth. After a few vertiginous seconds of falling, it disappeared, swallowed over the lip of the horizon. Somewhere a few miles away, on a silent patch of scrubland, the inert bomb plugged itself into sand and clay.

There was an awkward silent moment on the telemetry pad. Dick Johnson had his slide rule out, already poring over a strip chart coming from the high-speed camera, trying to divine if the test was successful from the little information he had: time of flight, how the bomb looked falling, how the pilot's voice sounded on the radio. Each team member offering a private prayer that the weapon had worked the way it was supposed to—and if it hadn't, that it was someone else's part that failed.

The range master gave the all clear, and the colonel sent every piece of earth-moving equipment on base out into the range to search for the prototype.

When they found the bomb, punched partway into the ground, it was a disappointing 148 feet from the target.

They built another prototype. They waited two weeks for another time slot on the test range, but when the time came, they set up the laser on the tripod, loaded the second prototype into the plane, and waited as the pilot narrated his approach.

Again the bomb separated from the plane and disappeared over the horizon. Again the range master cleared the range, the colonel ordered the earth movers out. This time they found the bomb plugged into the ground 78 feet away from the target. At first blush it looked like progress, but the data in the coffee can told a different story. The bomb hadn't actually steered. The fins had moved, but with little effect on the bomb's flight; the bomb had landed a little closer to the target out of luck.

The team decided they needed bigger fins. They waited another two weeks for test number three. This time, with the bigger fins, the bomb performed even worse. The data from the coffee can sensor revealed the bomb hadn't navigated at all, but now the problem was with the electronics: the Texas Instruments chip hadn't passed any information to the fins. The team was moving backward—and they were running out of money and time.

Another weeks-long wait for range time, a fourth test, another big miss. The TI laser-guided bomb was headed for failure.

Pilots over the Dragon would need a miracle to come from someplace. And if it didn't come soon, more of them would get shot down.

6.

Somewhere over North Vietnam
Goatee 01 Flight

Rick Hilton is in the sky, flying blind, somewhere above the Dragon. Finally sent against the thing itself, he's flirting with SAMs and 85's and the MiGs will be in any second. It's quiet so far, but he knows the quiet won't last.

He's straining to get a glimpse of the structure down below somewhere. That damned cloud deck moving under him again. His new back seater, Bill Wideman, is quiet. Hilton can picture Bill's eyes locked on the little five-inch Sony monitor bolted into the cockpit, watching a low-res readout of terrain scrolling by below.

"There's the mouth of the river down there," Bill says. "At about 11, I think. Got it breakin' through there?"

"Yeah." Hilton sees it. Sort of. A split-second gap in the clouds, and something black and shiny slips into view 22,000 feet below. "Well, I got *something* down there."

"It looks right to me," Bill says.

Bill's eager. Too eager. Hilton isn't sure, and this time it's not just him and Bill he's responsible for. There are sixteen planes on this mission and he's leading four of them. He needs to be sure.

The clouds swallow up the Dragon again and he's blind.

Hilton extracts data from the color of the sky. Another rapid calculation: humidity is dropping below dew point. He knows what that means. It's not just clouds; it's haze, as if even the sky is veiling itself, trying to hide the bridge. The enemy protected by guns, MiGs, and missiles, by American politicians who won't let them bomb the MiG airfields, and by nature now too.

Hilton hunts. "I don't—I don't see it."

"I'm sure it's on that side."

Hilton sees a dullish brown flash through the clouds, there and gone. "There's the river to the left."

"Yeah, that's what I'm saying." Bill's impatient in the back seat. "That's what I saw."

They're talking past each other now. Hilton has an urge for quiet. A voice crashes in from another plane, one of the parallel universes connected by radio.

"Got a bunch of triple-A this side of the bridge!"

"Which side is '*this side*'?!" But just as he asks, he has his answer: he sees the rounds start air-bursting all around him, the sky spotting with sudden plumes, and the mission begins to fall apart.

One of the planes has a flight control problem and has to bail. Their protection, code-named "Brenda" flight and "Bertha" flight, should've already dropped chaff upwind, a rain of thin metal to deflect radar beams, but they still haven't shown up. Without chaff, the planes will show up on enemy radar as clearly as they did the day he flew into the SAM trap. Only now he's above the Dragon, and he knows they have far more than just one weapon system down there hunting him.

He can't stay up here looking for the bridge forever, burning through fuel and waiting to be shot out of the sky. Every second above the Dragon is another chance for the bigger antiaircraft guns to find him, for SAMs to lock on, or the nimbler MiGs to take off. Fuel is dwindling and visibility is moving in the wrong direction. He feels the worry rising in other pilots, something passed along the airwaves as the sudden puffs of color keep flaring on either side of them, expanding discs of shrapnel getting too close.

Brenda and Bertha radio in. They're a full eight minutes behind schedule, an eternity. Another pilot makes the call: there's no time to wait.

"Confirm," Brenda radios back. "You *don't* want Brenda and Bertha in?"

"That's right, it's too late!" The pilot's voice is high and frantic, almost hysterical.

"Roger, making a port turn down here."

Somewhere out of sight, Hilton's protection pulls farther away. They'll have to continue the mission without chaff.

Down below, next to the Dragon, a seven-man crew rolls out a four-wheel trailer with a 57mm fast-firing antiaircraft cannon, a weapon capable of launching seventy rounds a minute. A radar array on top of a van scans the clouds and easily finds Hilton's formation orbiting above. The gun crew feeds the machine its sixty-three-pound clips, rotates the barrel up, and begins firing at the sky.

7.

Detachment 5
Eglin Air Force Base

In the hangar space the team had been given, Weldon looked over the next prototype. There was no real reason to believe this one would work. It was the same design as the last one. And the design, it was impossible to dispute, was ugly. There was little doubt the badminton birdie had been a stroke of genius, but it made the whole contraption look like some kind of misshapen submarine with a periscope permanently extended. Held out on a strut off the bottom of the bomb, the Giannini probe looked inelegant. It looked unnatural.

Weldon was facing the prospect of failure in his first important role. He was on the verge of letting down a company trying to spark a technological revolution with its new silicon chip. Worse, he was letting down young Americans flying dangerous missions over heavily defended targets. He could picture pilots over the bridge in the moonscape photo the colonel had shown him months before, flying their doomed missions.

He considered the design in front of him. Sitting still, it was ugly, but the more you pictured it in motion, actually moving through the air, the easier it was to imagine the inelegant design not just looking clumsy; you could see it causing problems. As the bomb fell, the force of the wind against that strut would be extreme, and if the bomb rotated in flight, the head would be whipped around violently. They'd been thinking of the badminton birdie as the head and the strut holding it as the neck, but maybe that was wrong. Maybe the strut was more like an arm. And once you began to look at it that way, the solution was almost obvious.

Weldon called the team together. They were going to try a redesign. All the same components, except they were getting rid of the strut. This time they'd attach the Giannini probe–inspired eye to the very front of the bomb.

They took their new design to the test range. They gathered at the t-pad again. They listened again to the pilot calling off his approach. They watched the bomb release, the crawling black speck. Again the bomb disappeared over the horizon and plugged itself into the sand, the range master gave the all clear, and the team went out to scour the ground.

Again they found the bomb embedded in the earth, only this time it was not even thirty feet from the target. For the first time the colonel showed excitement. They decided to test the same design again, and again they scored a direct hit: inside thirty feet of the target. The colonel was nearly ecstatic now and decided it was time to up the ante. He sent an old two-and-a-half-ton Army truck out into the range and had them aim the tripod-mounted laser at it. The colonel wanted to see the bomb actually hit something.

This time the test bomb landed twelve feet away.

The colonel upped the ante again. He wanted to see an even more real-world test. He had the team take the laser designator off the tripod, attach it to a rifle, and give it to a copilot riding in the back seat of an old Cessna. He wanted to see if the bomb still worked when the target was being illuminated by a laser shining from a moving plane. This time the test bomb landed within ten feet of the target.

Soon they'd run eight tests. The contract called for twelve, but the colonel was already sending excited messages to the Pentagon about progress down in Florida. Before their ninth test, he came to tell Weldon's team that the brass wanted to know if their laser-guided design would work with a bigger bomb. They went back to the shop, disassembled the final three prototypes, and built a custom kit for a single

3,000-pound bomb. This larger bomb test would have to be the last: they had no budget for anything more.

Before this final test, Dick Johnson had an idea. Given the mission inspiring this sprint for a new weapon, it would be fitting if the final target was a bridge. Out on the test range a crew dug a small pool and built a quaint little pedestrian bridge over it. As the team prepared the final test, the colonel sent Weldon up to Washington. There was no time to waste, no reason to delay even a day or two, and no more prototypes to build. Weldon landed, went through the usual security checks, and—just before meeting with Air Force officials—called down to Florida for news.

The team reported that the final test had gone perfectly. The bigger payload hadn't been a problem. The final bomb had landed within ten feet of dead center. The little pedestrian bridge hadn't stood a chance.

8.

Wright-Patterson Air Force Base

Weldon landed in Ohio soon after his meeting with Pentagon officials, having been given one more lobbying assignment.

Back at the Pentagon, senior officials were impressed, and there was now no denying the promise of a bomb guided by laser, but they still had a hard time picturing Weldon's little band of TI electronics nerds as real defense contractors. They wanted Weldon's team to wait so they could give North American Aviation another chance at the laser-guided bomb.

The colonel didn't have the patience for that. North American had already spent twice as much as Weldon's team and didn't have a single successful test to show for it. Who knew how long it would take for them to catch up, or if they even would? Why keep a potentially lifesaving weapon from young American warfighters any longer?

The colonel called some of his old friends at Wright-Patterson—which, as one of the biggest Air Force bases and headquarters of the Air Force Logistics Command, was the one place that had as many Air Force luminaries hanging around as the Pentagon—asked them to set a meeting, and dispatched Weldon to make his case before one more audience.

When Weldon walked into the Ohio conference room, there was one officer in particular who drew attention to himself, even sitting in silence. Chewing a cigar and leaning back with cowboy boots propped up on the table sat none other than General Curtis E. LeMay. Former head of Strategic Air Command, former chief of staff of the Air Force, perhaps the single most imposing figure in the whole young history of American airpower.

Weldon tried not to lose focus, but LeMay was impossible to ignore, both because he cut such an intimidating figure and because he was exactly the wrong audience for the pitch Weldon was there to make. LeMay was famous for shunning the notion of precision; during World War II he led a six-month sprint of purposefully indiscriminate firebombings over Japanese cities that killed hundreds of thousands of civilians and started city-sized infernos so intense, the tail gunners in the American planes reported being able to smell the burning flesh thousands of feet below. Perhaps the only act of discretion in LeMay's whole history of excess was when, late in the war, he sent planes to drop "LeMay leaflets" on Japanese cities, warning of the wholesale destruction the Allies had planned. "*We are determined to destroy all of the tools of the military clique . . .* ," the leaflets read, but warned that civilians were at risk, too, and should leave the cities because "*unfortunately, bombs have no eyes.*"

Now Weldon was presenting LeMay and a conference room full of Air Force aristocracy with a proposal for a bomb that did have eyes. But when he finished, there was little resistance. LeMay looked on. The only question the gathered generals seemed to care about was whether this idea, as promising as it sounded and as well as it had tested, would *really* be usable in practice. In *combat*. Weldon looked around the room. He made a show of hearing the concern and said he believed the answer was yes. Not only would it work, but, thanks to Jack Sickle, pilots and ground crews might find the new bomb surprisingly easy to use. And thanks to Dick Johnson's Giannini probe idea, it was going to be far less expensive and less susceptible to failure than the more complicated steerable bomb ideas the Air Force had already tried. By using laser designators attached to airplanes—an evolution from the colonel's test range idea of sticking the laser on a rifle barrel aimed from the backseat of a Cessna—you'd be able to both drop the bomb and also aim it from an airplane a few miles up and still strike a precise target with a decent chance of sparing whatever happened to be nearby.

LeMay puffed; he seemed to think; he shifted his boots on the table. In the end, he offered no challenge. The room approved, and the meeting was over in twenty minutes.

Now the colonel could say he had the backing of Air Force royalty. From there, the rest of the pieces were easy to put in place. The Aeronautical Systems Division set up a new office for Weldon's team called Project Paveway, diverted a half million dollars from other programs to this more promising one, and drew up a new contract for Texas Instruments to build fifty working laser-guided bomb kits.

As Weldon's team refined their design, preparing to graduate their invention from prototype into a production model, they decided to make one final, fateful change.

The clumsiest part of the design had always been the wiring, which ran all the way across the outside of the bomb, connecting the photon sensor and chips on the front—the eye and the brain—to the fins on the back. That had always seemed unavoidable. But as they prepared to scale up, Weldon began to see the unnecessary assumption they'd made: just because nearly every bomb ever built steered with fins on the back didn't mean there was some physical law dictating that rear fins had to steer. Even in the natural world, if you looked around, there were exceptions. Most fish used their tail fins only for propulsion; they *steered* with the fins on the front of their bodies. It was true that fins the bomb used to steer needed to be wired to its brain. Maybe it wasn't true that they needed to be all the way at the back. And that thinking unlocked another advantage. They could put all the key components—the side-mounted fins that shifted up and down to steer; the photon-sensing "eye"; the semiconductor chip brain—into a single piece bolted to the front of the bomb. The tail fins would just be there to stabilize the bomb in flight; they no longer had to move, which meant they no longer had to be wired to the front. It was a final adjustment that proved extraordinarily significant. It made manufacturing easier, it made the bomb more reliable, and it simplified assembly for the ground crews in Vietnam, who now wouldn't need to fiddle with wires at all.

More ideas emerged from Weldon's team, reinvigorated by the validation they were finally receiving. But before they could further perfect their invention, the Air Force called with an update: they were increasing their order from 50 units to 250. That brought all tinkering to a halt. It was time to start *building*. By then Weldon knew that when creative people were creating, it was best not to try to stop them. So, rather than ordering members of the team to stifle their ideas, he found an empty box, scribbled "creative backlog" on it, and told them to jot down any still-unexplored ideas in case there was a chance sometime down the line to modify the design. And with that, they got to work building the first working laser-guided bombs for the United States military.

The first reports coming back from Vietnam were nearly ecstatic. The pilots who dropped the first units were astounded. Analysts had to double-check their math. Officials reported that after the first few drops, bombing accuracy had increased by a factor of ten. They'd never seen anything like it. Memos back from the field described mission planners as "jubilant": officials had a hard time believing the new bomb so dramatically outperforming anything they'd ever seen before cost less than $6,000 per unit. It turned out that the bomb—which military officials started to call *Paveway*, after the Air Force office overseeing Weldon's team—had arrived just in time. President Johnson was in a predicament, unwilling to admit defeat in Vietnam but unable to accept the American death toll. He decided to start pulling American ground troops out of the country without declaring an end to the war. The military had to compensate for fewer boots on the ground with more bombs in the air.

Weldon got another call from the Air Force, increasing their order from 250 to 1,000. Then another, asking if the team could make 1,000 every month. Then 2,000 a month. Then, in the spring of 1972, three corps of North Vietnamese fighters launched a massive surprise operation, crossing into South Vietnam and seizing dozens of towns and villages. The South Vietnamese military was caught entirely unprepared, and by then fewer than 10,000 U.S. ground troops remained to help push back the advance or cut off supply lines. Without enough troops on the ground, President Johnson decided the best available response was a major increase in bombing.

American pilots went back up into the world's most dangerous airspace.

9.

Goatee 01 Flight
Somewhere over North Vietnam

Hilton catches a glimpse. "There's the bridge," he says. "There in the mist."

He's about to help Bill find it, too, when Bill starts yelling from the back seat.

"Triple A coming at us!"

Hilton identifies the caliber and type of antiaircraft artillery before he even realizes he has: this one's 57mm, seven-man crew, radar guided. He knew this mission would go sideways. They'd tried the Dragon's Jaw again a week ago, the first go at it with this new weapon that had just arrived in boxes on base. That mission failed: Hilton figured he must have banked too hard after dropping the bombs; his wing must have dipped into the laser beam and blocked it. The bombs lost guidance, went ballistic, and flew long into the river.

They'd accomplished nothing with the new weapon—*Paveway*, an on-base contractor called it—aside from likely rousing the enemy from whatever lull they might have been in. It took two weeks to get another mission ready, and now the enemy is surely waiting down there with their fingers on the triggers of every damned thing they have.

He's flying in formation again, four formations in all, and the only thing he has going for him is this new bomb that is supposed to free him from having to get down within a mile of the bridge like he has to with other bombs. The techs on base said that as far as they understood it, so long as a laser was shining on the target—so long as he didn't blind it with his wingtip again—he should be able to drop Paveway from three times as high.

The four formations bob in the air. He's waiting like he's always waiting: for a break in the clouds. Antiaircraft fire pops all around, but most of it is topping out a few hundred feet below him. The heavier guns will wake up any minute now. Never out of range of MiGs, and he's learned the hard way that the slyest SAM operators wait until you're at your most vulnerable.

He looks down again. He's lost the bridge again.

The leader of another formation settles into position and tips over into a dive, braving the 57mm and nosing toward the bridge, because even if this new bomb can really guide itself, it's still unpowered; it still needs to be dropped in the direction of the target. But the formation leader loses visual, aborts before weapon release, and radios out to the next formation: "If you have the target, you're cleared in."

The lead pilot in the second formation begins his dive, and this time the cloud break cooperates. Hilton watches the other planes pierce whorls in the clouds and stay on course long enough to release their bombs, then pull up before they get down into range of more antiaircraft fire. There's a long quiet, and then Hilton catches a glimpse of the bridge and sees bulbs of black smoke rising. It looks like some of the bombs must have landed on target, but then the smoke begins to thin. He can make out struts peeking through. The bridge still stands. *My turn.*

The plan is for Hilton in Goatee 01 to dive, drop his bombs, then pull up above. Hilton has three other planes following his lead, briefed to dive one at a time after him. Since only Goatee 01 has a laser designator, Hilton and Bill are supposed to keep orbiting the target from up high, continuing to mark the bridge with the laser while Goatee 02 dives, drops its bombs, and pulls up and out of the way so Goatee 03 can dive, then Goatee 04. Just before Hilton noses into his dive, the clouds begin to thicken again, whorls shifting in off the water. The bridge is starting to hide, and if he can't see it, he can't point the laser at it. He has to abort. But before he radios the other Goatee pilots, he catches a flicker of black through the clouds. There might be time for a dive; there's not time for four. Hilton improvises.

"OK, Goatee," he says, raising the whole formation at once. "Tighten it up. We're all coming in together."

In the sky, the formation changes, the four planes all turning their noses down at the same time. Together, they start to descend. Hilton is putting the men at even greater risk now. The planes have little room to maneuver without clipping one another, so they've just become one big, easy target. But their only chance is to all drop their bombs before the bridge disappears again. They pick up speed toward the ground. Bill starts talking again. "Beneath that cloud, right?"

He's looking back and forth from his cockpit window to the little five-inch Sony monitor he has back there.

Hilton's too busy flying; he's lost the target. "I can't see it. Can you?"

"I think that's it at nine o'clock."

"I know *where* it is. But I can just barely make it out."

A beat of dead air. Bill doesn't have much to add. "It's gonna be a bastard."

They punch through the cloud deck, picking up speed, the three other planes sticking to their wings. Hilton tries to ignore Bill, but Bill's still talking.

"Come on, bridge, where are ya? I saw it there out the window, but . . ." Bill needs to see the bridge on the TV monitor, too, or he won't be able to aim the laser, which is slaved to the television camera. Antiaircraft fire pops all around. A noise in Hilton's ear sounds as they dive, the plane squawking a warning he'll ignore for the moment. The MiGs should be taking off now, SAMs launching from their rails. A problem for later.

"There she is!" Bill screams, almost ecstatic. "Right . . . there!"

Hilton keeps steady, conscious only of how much altitude he's gobbling up.

"OK, I got it!" Bill speaking fast now, excited. "I can barely see it, but I got it!"

The plane hums. Wind scrapes the fuselage. Hilton trusts that the men flying alongside know him well enough to follow his lead. All he has to do is nod, and through the canopy glass the pilots in the four planes catch the shift of his helmet.

At the exact same moment, the pilot in each plane hits a button on his joystick. Fifteen Paveways tumble free from undercarriages. Thirty-seven thousand pounds of metal separating from planes and waiting for a signal telling it where to go.

Hilton feels his plane bounce up, suddenly lighter, but he has to stay in the dive. Behind him, Bill's been saving the laser bulb. Now that the bombs are away, with the bridge in his crosshairs, he flips the button that turns the laser on.

Below, each bomb wakes up. Each sees the light reflecting off the bridge, registers energy in a section of its eye, releases electrons into TI's semiconductor chip, which sends a signal to the gas canister, which triggers a blast of gas that moves the fins. All fifteen bombs turn toward the target. Just as they begin to overshoot, the fins kick the other way, all of it happening in milliseconds—the bomb swimming through the air, a zigzag course following a flashlight only they can see.

Hilton starts pulling the plane out of the dive. But as he pulls the nose up, he's pulling the laser beam up too. In the air, the fifteen falling bombs follow the laser reflection as it slides off the bridge and out into the water.

"Easy on the pull!" Bill's yelling. "You're pulling me off!"

"I'm trying to hold it!"

As Hilton pulls up, the force of gravity is crashing down. Lifting his hand is like lifting a weight three times heavier than his actual hand, four times, and in the back seat Bill's on a joystick with a hand that's getting heavier and heavier as he tries to get the crosshairs back on the bridge.

Bill lets out a groan like he's in pain. "Ahh, I can't! I can't!"

Hilton glances at the altimeter: 10,000 feet, nearly two miles above the earth, and coming into range now of more antiaircraft weapons. He makes another flash decision: mission first, survival second. MiGs, SAMs, triple A—*Come and get me.* The only thing he'll pull up to avoid is the ground. He eases the stick forward again, slacks off the g's, and lets the nose settle back down into the dive.

Below, the bombs see the sparkle slide back up out of the water and onto the bridge. They correct course, each bomb whipping its fins back and forth, up and down, trying to center the reflection.

"OK," Bill says, "I think I got it back. I'm not sure."

Hilton can feel the frenzied energy back there.

"OK," Bill says again. "There's an impact."

Hilton flexes his fingers around the stick. It takes a moment, but then he sees it, too: flashes off the ground. Now he yanks the stick back out of the dive, letting the g-forces multiply because Bill's work is done. The bridge cloaks itself again, this time in smoke and debris, and the men again lose their view of the ground. Hilton pulls the nose higher and higher, bracing for the plane to start screaming there's a missile on his tail, or for Bill to call out about MiGs or 37mm. He keeps his breath steady to ready himself for evasive maneuvers.

But for a moment, with the Dragon hidden, it's quiet. No sound at all from any of the planes. Just jet whine and wind. The weapons below haven't woken up. Even the 57mm has tapered off.

"God," Hilton says. Even for Hilton, the silence and calm starts to get uncomfortable. "It was almost impossible to see in that haze."

"Yeah, very tough, but I had it."

More silence. Something's changed in the plane. Suddenly they're both excited, talking over each other.

"Until—"

"When you said '*I had it*'—"

"Until you don't—"

"I hit—"

They both shut up. Something's become uneven.

"When you said *I got it*, I looked down," Hilton says. "We were going through 10,000 feet, and I was being as gentle as I could."

"Yeah, well, I got it back just in time for explosion. I don't—I don't know. You wanna take another orbit around it?"

Extra time above the Dragon is a bad idea. "Yeah, might as well look at it."

Bill lets out a little giddy laugh. "*Eee*-aww!" There's a chance they've made history. "Well, there's a hell of a lot of smoke down there."

"Sure is."

A lot of smoke, Hilton thinks, and surprisingly little antiaircraft fire. For the first time in a while, he's confused. Something isn't right. It feels like there's a glitch in the system. The stitch that holds the universe together virtually undefended.

They make an orbit and still no noise from below, no warning lights in the cockpit. Hilton sets a heading for Ubon and spends the rest of the flight trying to sort out the one equation he isn't able to effortlessly resolve. Even after they land, he still can't quite square it. The mission was easy, almost uncontested. It's strange.

Later, after he's showered and changed, Hilton retires to his room. He's recording a new tape for his wife when the phone rings. He apologizes to Ann as if she's sitting there with him. He picks up the phone, and it's the captain calling to give him the news.

The Dragon's Jaw has fallen.

BOOK III

WIEGAND

(OPERATION EL DORADO CANYON)

1.

Building 2
Ministry of State Security ("Stasi") Headquarters
East Berlin

From his office window Rainer Wiegand, veteran counterespionage officer, loyal servant to the Fatherland—at least for a while, at least until recently—could look down and see meaning.

Wiegand was nearing the twilight of his career. Nearing, if he was honest, the end of his career.

He was more inclined to notice things like this now. There were violent threats out there, ones he should be chasing, fighting, threats he should be preventing. But his superiors wanted him focused solely on his own people. He saw his ministry in broadening light. Its disregard for foreign threats to its people. Its interest in oppressing them instead. He could see, four stories below him, a rack of squat houses, squeezed together and subdivided. Crammed with families living in the actual shadow cast by the massive headquarters of the agency Wiegand had spent decades serving. He could see this now. The unhidden symbolism of a regime he'd spent his adult life defending. This building, his building, which dominated the space. It filled up the horizon, so its power was apparent in every moment of every life as if designed for that reason. Fear deters disloyalty. Deterring disloyalty makes the job of finding and punishing it a little easier.

He could see he was beginning to think a little too independently for his own good.

He wanted to be chasing foreign spies, and instead the Stasi's obsession was studying its own people for seeds of treason. Sniffing around families like the ones slummed into those bisected and trisected little hovels, writing them up for petty offenses and adding them to lists.

Wiegand would just as soon leave them alone. He had little interest in turning Germans. His innocent countrymen, just civilians. He was having a harder time pretending he did. They were not the problem.

The real problem, Wiegand believed, was the Arabs. Mostly the Libyans. One Libyan in particular. A young man named Eter who arrived in Germany under diplomatic cover.

From surveillance photos taken at the airport, you could tell. Eter arrived in tight-fitting jeans that flared out at the bottoms, hair in a picked-out Afro. Skin just dark enough to seem a little exotic almost everywhere, completely out of place almost nowhere. Of all the officials in all the surveillance photos Wiegand saw, something about the one called Eter rubbed him the wrong way.

A file search in the new computer system confirmed easily enough that this new arrival wasn't a diplomat. The Libyans were either not very good at building legends for their operatives or they just didn't care much about keeping them undercover.

Without much effort, Wiegand discovered that Eter—full name Musbah Abdulgasem Eter—was not only part of Libya's spy agency; he belonged to a special unit tasked with running operations abroad. Eter was a militant.

Wiegand leaned in. He studied the embassy. It wasn't just Eter. Too much activity, a sudden staff expansion. That didn't make sense. It wasn't like the Libyans suddenly had more diplomatic work to do. Something sinister was happening at the embassy, and Eter was some kind of leader. You could almost see it in the way he carried himself. Even in still surveillance photos, the young man had a swagger.

Wiegand gamed it out. If Libyan militants wanted to operate in Europe, they'd need a forward operating base somewhere. But western Europe was cutting ties with Libya, expelling their diplomats and forcing embassies to close because they saw what Wiegand saw and his bosses didn't: that Libya was becoming a terrorist state.

Most of Europe saw it, but Wiegand's own East Germany didn't. If Libya had an agenda in Europe, this might be just about the only place they could operate from.

As Wiegand watched, he began to see this Musbah Eter as a key node: the main contact point between the Libyan spy agency's headquarters back in Tripoli and a group of officers—terrorists, Wiegand was sure—operating right under his nose. Probably planning some kind of savage attack on a target right here in Europe.

2.

Northwest Texas

At Texas Instruments, business was good. The undroppable bridge had been dropped, one run with the new smart bomb achieving what seven years of missions before it hadn't. Soon after the Dragon's Jaw, a mission against five more bridges in North Vietnam succeeded in complete destruction of their targets with no lost pilots, no damage to planes, and only twenty-four bombs used. Air Force planners calculated it would have taken 2,400 regular bombs to achieve the same effect before Paveway.

Paveway solved the tactical problem pilots faced, but also the political one Washington did. Washington was suddenly comfortable letting war planners dispatch bombers against SAM sites and MiGs parked on the ground, because now they could be hit with reliable accuracy, and collateral damage to any Chinese or Russian advisors or anyone else who happened to be nearby could be avoided with equally reliable consistency.

Reports from early pilots over North Vietnam describing Paveway as almost eerily easy to use began to spur more demand, and the bomb began to argue for itself. Missions expanded. Paveway spread. The Air Force deployed the bomb all over North Vietnam. In a single twelve-week sprint, one hundred bridges were destroyed. Paveways were dropped on ports without concern for flags from other countries and on power plants without concern for civilian contractors nearby. Bombing performance reversed, from missing almost 90 percent of targets to hitting almost 90 percent of targets. Air Force researchers spent the last months of the war scribbling out their findings and running the numbers again, wondering whether the new smart bomb from Texas Instrument's team of unknowns could actually be *this* good. Against hard targets, Paveway proved two hundred times more effective than conventional bombs, and of the thousands of Paveways dropped during the last months of the Vietnam War, more than a third were recorded as landing within a few yards of their mark. Half were direct hits. In four months, the United States destroyed more targets than it had in four years. By the end of the war, Texas Instruments had delivered 100,000 Paveways, ushered in a revolution in warfare, and achieved their original business goal of helping to spark a market for the chips used not only in Paveway but also in a growing line of devices capable of basic "thinking."

It was a change of a generation—or maybe two or three—within the closing chapter of a single conflict. It was almost just a bonus that this new weapon proving so revolutionary in the field was also so obscenely cheap and so easy to use. Jack Sickle's mantra of simplicity had the intended effect: ground crews liked Paveway; pilots wanted more of them.

Texas Instruments was ramping up Paveway production so quickly, it ran out of space, and soon after its creation Paveway needed a new home. That presented a problem: Paveway was a weapon built by a company that wasn't really a weapons company. The only available space TI had was being used for their electronics work, a North Texas plant where TI had begun to manufacture the new semiconductor chips. But it turned out the highly particular specifications required for a chip plant were not so different from what you'd want for sensitive military tech: careful monitoring of who and what went in and out and a sterile, well-controlled environment. TI decided the chip plant might in fact be an ideal place for Paveway and began shifting assembly for America's new favorite bomb to the same building where its brain came together.

As Weldon charted a course for his new product, TI set out to find the best people to assemble it, and for that, too, the company's other electronics work proved useful.

Working the Paveway assembly line could be dangerous, even for those who had nothing to do with the actual warhead. The special batteries that powered the guidance unit gave off corrosive chemicals, and the canisters full of pressurized gas that moved the fins could explode.

Being as discerning as possible about how they staged the assembly line could be a way to minimize mistakes and injuries and maximize efficiency. The company had been making chips for long enough now that they'd begun to see that when it came to the kind of fine motor skills and steady hand required to do things like weld connector wires under microscopes, it often seemed like the best workers were young women.

There was ongoing scholarship on factory work being conducted in the United States and overseas, much of it confirming what TI was beginning to see in-house. There were studies suggesting that women had a lower median hand tremor; that with training, women were better than men at improving speed and precision; and that the age at which that kind of coordination peaked was about eighteen.

To Weldon's Paveway team at Texas Instruments, what it all meant together was that the ideal assembly-line worker for Paveway was often a young woman. The company sent out recruiters all over the northeast corner of Texas, and soon Paveways started rolling off an assembly line manned increasingly by young women. Demand for Paveway was continuing to grow, first in America and then abroad, and soon TI needed more girls than northeast Texas had to offer. When the company set out to find a location for its first international factory, it settled on the outwardly baffling choice of Bedford, county of Bedfordshire, middle of nowhere, England. The sleepy little shire offered very little that might interest an up-and-coming defense business—with one exception, invisible unless you were looking for it: the site on which TI broke ground for its new factory was situated within a mile and a half of two girls' boarding schools and an orphanage.

Along with a new influx of women assembling Paveway and other electronics in both Texas and England, the bomb was becoming even more reliable—and, for TI, even more cost-efficient. Competitors tried to break into the market with other ideas for "smart" bombs, but their products didn't work as well and often cost five times as much. Others tested impressive new capabilities, but Weldon remained steadfast that TI not try to outdo them. Paveway had to remain simple, unflashy, easy to use. It had to stay cheap.

"This needs to be about as much as a low-cost family sedan," Weldon said again and again, both to executives trying to beat competitors with new capabilities and to new engineers joining the team. Coming from someone more prone to listening than speaking, the consistency with which Weldon preached his "low-cost family sedan" mantra penetrated. It was sometimes hard for colleagues not to picture a Chevy Caprice tumbling out of a plane. "It's not how much can I get for one," Weldon said, "but how *many* can I sell?"

There were other ways the company kept the cost down. When Weldon first landed at Eglin, TI's on-base reputation as a band of electronics nerds had been a handicap. But now the fact that TI built both the bomb's body and its brain served to steepen its price advantage. And as TI looked for other ways to sell its chips—the first handheld calculators, which would become ubiquitous among high school students; word processors for corporate executives—it was building a virtuous cycle in which American schoolkids and office workers were, unbeknownst to them, helping to realize the dream of a perfect, affordable precision bomb.

So, in the years following the chaos and human loss of an ugly war in Vietnam, there appeared to be at least one silver lining: the United States had a new weapon. It was cheap, easy to use, helped pilots stay safer, and could strike small targets while sparing people or buildings nearby. It could be deployed even if you weren't in all-out war. It could be used even if you weren't sure you wanted to go to war at all.

3.

The White House

Ronald Reagan wasn't sure he wanted to go to war, but he had a problem he didn't know how to solve.

He'd inherited a Cold War already seven presidents old, now fused to another, newer threat. International terrorism was a rising menace, frightening but ill-defined violence about which he understood little except that it often seemed to originate in one faraway country: Libya.

Libya seemed to be lurking behind all the most horrific acts of terrorism. After members of a death squad attacked the Israeli Olympic team at the Munich Olympics in '72, killing eleven of them as well as a police officer, investigators said Libya had provided the terrorists with diplomatic pouches to smuggle the weapons in. In '73, Libya was linked to a terrorist attack that killed an American ambassador and one of his deputies in Sudan; in '75, when militants stormed a Vienna OPEC meeting and took more than forty hostages, investigators again pointed the finger at Libya. And now the CIA was telling him that a hit team dispatched by Libya had just crossed into the United States from Canada on a mission to assassinate him and top members of his national security team. Now they needed to begin taking all sorts of inconvenient measures. Thanks to Libya, Reagan had to start wearing a bulletproof vest.

Reagan believed he was facing a malign actor operating with impunity, a metastasizing evil that had to be stopped. But how? He had the germ of a concept for economic pressure. He had a press plan his comms people came up with to help get the papers lined up against Libya. And that was more or less it.

He couldn't be sure the United States' allies shared his concern. The moral stain of America's involvement in Vietnam hadn't faded, and Reagan didn't think he could get foreign countries on board for some kind of big, faraway war. Or even his own Congress, for that matter.

Even putting all that aside, Reagan's America didn't use force based on hunches. Libya tended to leave partial fingerprints on all its plots, always enough evidence to suggest their involvement but never quite enough to prove it.

And when it came to using force, it wasn't clear what was legal. Arguments in the Situation Room spun around and circled back on themselves. National Security Council members wound up switching their typical positions. The country's top warrior, Secretary of Defense Caspar "Cap" Weinberger, found himself holding out for diplomacy.

Secretary of State George Shultz, the nation's top diplomat, pushed ever more boorishly toward war.

"With all due respect to the technical points," Secretary of State Shultz yelled as an overlong meeting began to strain his patience, "the problem of terrorism is *upon* us! . . . At some point, we have to wrestle with the issue of how 'sure' we have to be before we hit back!"

The issue that kept getting in the way of decisive action was the question of proof. What proof would the United States need in order to justify military action against another country? As far as Shultz was concerned, they all knew who was behind most of the mischief. Why wait for corroborating evidence while more and more people died?

But for Secretary of Defense Weinberger it wasn't only a question of *how sure*; it was also a question of *how*. Shultz wanted to "hit back," but there was no obvious way to strike a distant country without some kind of extensive military mobilization, and America's recent history didn't inspire much confidence. It wasn't just Vietnam. The last military operation overseas was the disastrous Operation Eagle Claw: a joint Navy, Air Force, and Army mission intended to rescue American hostages in

Iran, and which instead resulted in two U.S. aircraft crashing into each other in the Iranian desert and eight dead American servicemen.

General John W. Vessey Jr., chairman of the Joint Chiefs, disagreed with Weinberger's position. "The JCS have the same concerns as Secretary Shultz. We need to face up to the issue of state-sponsored terrorism," General Vessey said. "We need to decide that the next time a terrorist acts, we will hit back."

William Casey, the director of the CIA, spent the meeting wondering openly whether the United States even had the ability to wage the kind of mission they were now talking about. What equipment would it require? What weapons? A transcontinental attack against a distant target like Libya would surely mean moving military hardware across other countries' territory.

"What we need," Casey mused out loud, "is the kind of international cooperation that we had in the 18th century in the international laws against piracy. There has to be a consensus among the Western powers that they will all participate," Casey said, because, with cooperation like that, "operational capabilities are less important."

But that was just it. Western powers were not in consensus. America was not in consensus. This one room was not in consensus. Without international cooperation, "operational capabilities" would still be important. Without cooperation, the only way would be some solution provided by the equipment itself.

Reagan's National Security Council was at an impasse. Just about the only thing they could agree on was that they weren't agreeing on very much at all and that they ought to at least establish the terms of the debate. The council members came out of that meeting concurring that they needed a document that might at least make future meetings marginally more productive—a document that would lay out how the United States might punish countries involved in terrorism, or perhaps even stop them from sponsoring terrorism in the first place. A doctrine not just responsive, not just punitive, but also preventative. *Preemptive.*

They'd call it "preemptive" strike.

4.

Building 2
Ministry of State Security HQ
East Berlin

Lieutenant Colonel Rainer Wiegand studied the Libyans, unaware that his drift toward a kind of perilous independence was happening at the most critical time.

The Cold War would be over soon, but you couldn't tell here at its biggest fault line.

To the spies still plying their trade in the grayed, graffitied urban trenches, the Cold War was still at its coldest, which meant its hottest. Berlin was still the city of perpetual espionage, as John le Carré called it, divided up like a chessboard for a neatly cleaved world.

If, across the wall from Wiegand, West Berlin was a Babylon with its ostentatious cult of openness, the East was a city-sized prison. There were 200,000 political prisoners in East Berlin, a meaningful part of the population doing time for the crime of trying to get out, or even just wanting to. Recording devices were hidden in coffeepots and light switches. Beneath Wiegand's office, windowless trucks with logos for laundry services circled the city to confuse the political prisoners shackled inside. Scent samples from citizens of interest were held in front of sniffing dogs while machines designed to open, replicate, and then reseal envelopes zipped through nearly 100,000 pieces of personal mail a day: spying on an industrial scale, the whole city wired to hear. Beyond the Stasi's 90,000-strong employee roll, 175,000 otherwise normal civilians served as official informants, and at least half a million had been compelled to collaborate in one way or another. Husbands spied on wives, employees spied on bosses, prisoners on prison guards. By proportion of the population, Wiegand's employer was the biggest spy service ever built in human history.

West Berlin did things differently.

In the West they let people move with obscene easiness. West Berlin made a show of its cool. Up on Devil's Mountain, the city's highest real estate, a set of giant white domes hung as a taunt. Radars and antennae behind white panels of fabric, inflated by internal air compressors. The big round radomes of an NSA listening tower, built on top of broomed-together World War II rubble. The West didn't bother to hide it. The West flashed its prowess so you could see it from the East. The giant white domes almost indecent-seeming, the veiny genitalia of some albino giant, so people took to calling the listening station the "Berlin Balls."

And because West Berlin was a theater of people freely coming and going, it was a haven for migrants from all over the world. The West flaunted that too. Migrants came and mostly assimilated into the half city priding itself on openness, but some unsavory types slipped into the stream, too, silt driven in to scrape up against the melting pot, and there were causes to go around. Communist versus capitalist, East versus West, but beyond that too. Berlin was a theater in miniature for the cold parts of a dozen other wars. Arabs spying on Jews, Jews on Arabs, Arabs on other kinds of Arabs. Undeclared representatives from various liberation movements. Nationalist movements, extremist movements, Marxists, anti-Marxists, Stalinists, all dropping into Wiegand's view to feed on the carcass of a city pinned down and suffocating under ninety-six miles of wire mesh and concrete.

In Berlin, two tectonic forces had shifted into each other and driven up a wall that had split the city in half, not just the city but the country; the world, even. The front line in a Cold War that was staring to throw off sparks—and still, amid all of it, Wiegand believed he knew where to focus.

The Libyans were planning something, this Musbah Eter was planning something; Wiegand could feel it. It would happen somewhere nearby. It would happen soon. His colleagues tried to speak sense into him—*The Libyans are our friends!*—but he couldn't let it go. He was obsessed. The Arabs were scheming right under his nose; he knew it.

He needed human intel to prove the Libyans were terrorists; to uncover their plot. He needed a spy with access to the Libyan embassy—someone who could get inside the building and close to Musbah Eter.

He knew he was going out on a limb. Running against his bosses was running against the world's biggest spy agency. Still, he pored over surveillance reports, looking for hints of a weakness, dirt, something embarrassing he could use as leverage.

He found it quickly. Easily, even for a veteran like him: a Libyan official who, though he wasn't on Eter's team, worked inside the embassy and had left an almost comical trail of indiscretions: easily traced currency trades on the black market; visible, public drunkenness; poorly concealed visits to prostitutes, almost like he wanted to be caught. Wiegand's approach went seamlessly.

Intimidating the Libyan was easy; the overture went smoothly. Wiegand made his offer. *Silence about your misdeeds in exchange for your service as an informant.* The Libyan agreed with little fuss. Wiegand had his mole.

He arranged a first interview at a secret Stasi safe house, where he set up a tape recorder. If the interview turned up any details, he'd have proof.

The new mole, code-named "Mustafa," hardly needed warming up. That first safe-house rendezvous was a bonanza. Mustafa confirmed all Wiegand's suspicions on tape: he admitted the Libyans were using the embassy as a base to plan terror attacks and that Eter, along with a small team of operatives, was indeed working on a plot against the West. Mustafa listed the secret agents operating under cover at the Libyan People's Bureau and then, toward the end of the meeting, shared one more chilling detail: inside the embassy he'd seen with his own eyes a suitcase full of weapons. Pistols and silencers for sure and a package he thought could be explosives.

Wiegand knew what that meant. An attack was imminent.

Wiegand left the briefing and got in his car, using all the restraint he could muster not to speed directly back to his office. He forced himself to practice responsible tradecraft, looping around and reversing course to check for tails and prevent other spies from connecting the secret safe house to the Stasi.

He got back to HQ and carried the tapes inside. He couldn't risk anyone seeing them yet. Informants were all over. His own secretary could be spying on him. The janitor, a visitor, anyone. He knew how his bosses worked, how the head of the Stasi, Erich Mielke, worked, and Wiegand had just directly contradicted orders. He'd done right, he'd done good, but if he didn't handle these next steps delicately, he'd be the one shackled in a windowless truck with laundry service logos, being spirited to some underground prison, taken off the board before he had the chance to stop whatever bloodletting the Arabs had planned. He dismissed his secretary and sat down to type out a transcript of the interview. He worked without pause, finished the Mustafa transcript, marking it TOP SECRET and SPECIAL SOURCE PROTECTION and, before running it up the flagpole, summoning a second opinion from someone he thought he could trust.

Wiegand's friend Wolfgang Stuchly was the lieutenant colonel in charge of Section 15, the Stasi unit overseeing protection of embassies. He had relevant experience, he was at the same stage of his career as Wiegand, and they'd been close colleagues for nearly twenty years. Wiegand felt he was the right person to bring into the fold—that he could at least rely on Stuchly not to inform on him.

When Stuchly arrived at his office, Wiegand turned the radio up and gave a full briefing of what he'd been up to and then showed Stuchly the Mustafa transcript.

Stuchly read through it and looked up at Wiegand. He believed it, he said. All of it. And Stuchly in fact was running his own undercover asset, a Palestinian-born informant named Ali Chanaa, code name "Alba," whose intelligence matched Mustafa's. Not only that: Alba was getting close to Musbah Eter, the same stylish Libyan spy Wiegand was now most preoccupied with.

Even so, Stuchly begged him to back off. Their bosses were aligned with the Arabs. Minister Mielke had personally arranged training for Arab extremists; he'd helped supply them with weapons. Wiegand was not just violating the boss's orders; he was going against stated, demonstrated Stasi policy.

Wiegand boiled over. "I cannot allow those Arabs to blow people up!" He knew about Stasi policy, he bellowed; he didn't need this from someone he trusted. He didn't even care who the Libyans were plotting against, he told his friend. "We are not murderers!"

It was a dangerous admission, more irresponsible still for having been made inside Stasi HQ, where the walls listened even more ardently than in the rest of Berlin. Whether impressed by Wiegand's passion or just realizing he wasn't going to change Wiegand's mind and wishing to end the conversation before it got any more dangerous, Stuchly relented. Wiegand laid out his idea: if they could just show their superiors proof of terrorist activity under their noses, they'd be compelled to help stop it. With a new half-willing ally, Wiegand wrote up a report about an imminent attack, attached it to the Mustafa file, sent it up the chain of command, and hoped the Libyans wouldn't launch whatever devilish operation they had planned while the Stasi dallied over the paperwork.

When word came back down, it was perhaps what Wiegand should have expected. Rather than grant approval to expand surveillance of the Libyans, they accused Wiegand of being a bully. Libya was a poor country, defenseless against America and its powerful allies. They warned him to leave the Libyans alone; the last thing they needed was Wiegand adding to their troubles as they tried to run the one embassy they still had in Europe.

Wiegand fumed. *They're blinded by Stalinism!* He called Stuchly to rant: The Libyans weren't innocent; they had to know that. They just didn't care about whoever the Libyans' victims might be because the bosses assumed it wouldn't be *them*.

Wiegand decided to ignore their orders. He'd go a step further than just surveillance of the Libyans. He'd initiate a maneuver that bordered on treason. If he wasn't getting any help from his own agency, he might need the help of another one.

He gathered some of his best confidential informants, mostly Arab journalists, and sent them across the wall to West Berlin with instructions to begin frequenting the American press clubs there and to show up at Western embassy events. Wiegand told them to talk a little too loudly about their frustrations with the communist regime in the East, and he provided each of them morsels of information he knew would be of interest to the class of professional listeners always at the next dinner table over. One by one, the informants reported back. They'd each been approached by friendly men and women with American accents who happened to be at the same party, the same restaurant, who'd sidled up and initiated conversations revealing themselves to be interested—rather intrigued, in fact—by the opinions overheard, and would the journalists perhaps be interested in some kind of arrangement?

Wiegand had baited the hook; the fish had bitten. He wouldn't use it quite yet, but if he needed to transmit a message later, he now had a secret line of communication: a small stable of double agents newly recruited by American intelligence officers.

5.

Situation Room
The White House

Almost no one outside knew it, but inside the White House Situation Room, a quiet revolution was taking place.

National Security staff finished drawing up a blueprint that might allow Reagan's divided advisors to agree at least on a *theory* of preemptive strike. What would justify an attack against a country like Libya? What would the goal of such an attack be?

The top secret document, called National Security Decision Directive 138, said that whenever "we have evidence that a state is mounting or intends to conduct an act of terrorism against us, we have a responsibility to take measures to protect our citizens, property, and interests." The directive further ordered the military to develop a "strategy that is supportive of an active, preventive program to combat state-sponsored terrorism before the terrorists can initiate hostile acts."

Now the United States had a legal framework to launch attacks against other countries even if America was not at war with them.

Two weeks after Reagan signed the new, secret directive, Libyan leader Muammar al-Gaddafi struck again. A gunman inside the Libyan embassy in London opened fire on protestors outside, injuring ten and killing a police officer trying to manage the crowd, while back in Libya, Gaddafi sent soldiers to surround the British embassy and hold its staff hostage. To Reagan, this was exactly the kind of blatant, unprovoked attack against Western interests the new directive compelled a response to.

But Secretary of Defense Weinberger objected. Aside from the question of how exactly to carry out an attack in Libya with no pre-positioned forces—the "military strategy" NSDD 138 ordered but which hadn't been produced yet—an operation against Libya would actually not be a proportionate response, for one important reason: the Libyan attack had not been against American citizens. It wasn't just a question of legality. If America deliberately attacked Libya now, it would be an escalation, not a reciprocation.

Reagan found himself talked down for the moment, if just barely. He'd hold back for now, but his restraint was fraying.

Soon NSDD 138 was tested again. At approximately 9:10 a.m. local time on December 27, 1985, four men in blue jeans walked into the departure hall of Leonardo da Vinci International Airport in Rome carrying Kalashnikovs and grenades. They approached the desk of El Al Airlines, the national carrier of Israel, and opened fire. At the exact same time, next door in Austria, a nearly identical attack was underway at the Vienna International Airport departure hall. By the time authorities at the two airports had managed to take back control, the attackers had injured over a hundred people and killed sixteen—among them, four American citizens.

When the National Security Council met, Reagan was unambiguous. "I think we're all agreed," he said, "we must do something in view of the massacres in the airports at Rome and Vienna."

They already had proof of Libya's involvement. Investigators had quickly traced the attackers' weapons and passports to Libya, matched their grenades to a Libyan military stockpile, and secured reports from intelligence sources suggesting a money transfer from Libya to the extremist group whose men had carried out the attack. This time it seemed almost irrefutable that the blood of the innocent civilians was on Gadafi's hands. And this time, unlike his attack against the British embassy, the blood of American citizens was on his hands too. NSDD 138 was completed, signed, and in effect, and now, outside the Situation Room, the broader news-consuming world was beginning to wonder what America was going to do about Libya. The major newspapers ran three-column-wide photos of the carnage at the airports, with above-the-fold banner headlines. The rising scourge of terror was now impossible to ignore. Legal justification was in place. Libya had to be attacked.

The only argument available to those like Secretary of Defense Weinberger who'd campaigned for restraint was even more brittle than the one he'd used last time: Yes, Americans had been killed, and yes, all signs point to Libya. But the Americans at the airports in Rome and Vienna had been in the wrong place at the wrong time. The attack seemed to be targeting Israeli interests, not American ones. And since there was no proof that Libya was specifically, deliberately targeting the Americans, an attack deliberately targeting Libya would—technically—be disproportionate.

Even more reluctantly this time, Reagan acquiesced. It felt like half the NSC staff was braying for military action against Libya, but for now Reagan would take every step just short of war available to him: a trade embargo against Libya; an impressive, intimidating military exercise off the Libyan coast. And in the meantime he wanted the military solving the problem of operational capabilities: how to go about launching a long-range attack, with limited support from allies, against a dangerous adversary on foreign soil. He wanted the plans on his desk and ready to go.

The Libyans had done perhaps the worst they could do without triggering an American response. The only trespass left—perhaps the only line Gaddafi hadn't crossed—was an attack that killed Americans on purpose.

6.

Building 2
Ministry of State Security HQ
East Berlin

New threat intel landed on Wiegand's desk almost every day now. Intercepted cable traffic grew more worrying, and activity at the embassy increased after American forces taking part in a military exercise off the coast of Libya crossed into waters Gaddafi had claimed as off-limits and then, when Libyan forces engaged, the Americans responded by sinking three Libyan vessels, killing twenty-four Libyans. The Libyans were riled up and seemed to be increasingly angry at America.

Wiegand read surveillance reports and translated intercepts that together showed what looked like an unmistakable move toward some kind of operation. Reports that Musbah Eter, the object of Wiegand's obsession, was spotted casing locations in West Berlin; reports that he was sending ciphered telexes about potential targets back to his HQ in Libya; that explosives had arrived in East Berlin and were being stored somewhere in the Libyan embassy. And Wiegand's one reluctant Stasi ally, Lieutenant Colonel Stuchly, said his confidential informant Alba was reporting on plans he'd overheard the Libyans making.

Wiegand was getting so much credible intelligence on potential Libyan plots that it was getting hard to keep them all straight. A plot to use grenades and handguns to attack a bus that carried American soldiers from East to West Berlin every day; a plot against a bus that carried the children of American diplomats; one to detonate a car bomb in front of the pro-Israel Springer Publishing Company.

Libyans were spotted around West Berlin; they seemed to be casing places with obvious American connections. The U.S. Army's hospital near the Berlin Botanical Gardens, a military gas station, the U.S. Army Barracks, a drive-in restaurant frequented by Americans, a nightclub called Nashville in the American sector of West Berlin, another called La Belle, one called Stardust. Alba reported having himself driven Eter to the American consulate general and to an American movie theater called OutPost. It was like a lever had been thrown and the flood of threat intel was coming in too fast to process. The Libyans seemed to be considering attacks against every location that had any American association at all. That, or they knew they were being watched and were trying to make it more difficult for the watchers.

Either way, Wiegand was hamstrung. Aside from Stuchly, he had no support from other Stasi departments and knew he had to avoid attention from his superiors. He was having a hard time deciding where to focus his limited resources, didn't have the manpower to investigate every potential plot, and had no way of knowing which information to ignore and which to focus on—until he saw one intercepted communiqué between Eter's team and their headquarters in Libya that contained something none of the previous messages had: a date. An action was to be taken on March 26.

Wiegand grew frantic. He felt he had no choice but to try once more to get his superiors to help. The response this time was even worse. Wiegand was ordered to immediately cease his surveillance of the Libyans. He was expressly contradicting explicit orders to leave them alone, and he was told he'd be held personally accountable for any damage to the relationship with Libya.

Wiegand fumed. He brought Stuchly back in, who nodded along as Wiegand railed against the "comrades of ours who shut their eyes," Stasi operatives who were fine with civilians being killed so long as they weren't German. The Arabs were wily and self-serving; their loyalties could easily change, and the Stasi was playing with fire. "They could turn against *us* tomorrow," he said. "*We* could be blown up just like anybody else."

Wiegand didn't need to tell his friend he had no intention of following the orders. He'd dispatch everyone under his command to stop the Libyans' March 26 plot and beseeched Stuchly to provide anyone he could muster too.

THE DAY BEFORE the expected attack, Wiegand's men reported movement.

Two employees from the Libyan embassy getting into a Volkswagen Golf with East German diplomatic plates, then driving toward the Berlin Wall.

A little after noon, the car successfully passed through Checkpoint Charlie and entered West Berlin.

Then, once across the wall, something curious: rather than proceeding toward the Springer Publishing Company's building or any of the other potential targets, the Volkswagen turned in to the Kreuzberg neighborhood, one of West Berlin's poorer areas, and pulled up to a housing complex.

One of the men got out of the car carrying a suitcase—a suitcase that matched the description of the one Wiegand's informant Mustafa had reported seeing packed with silencers, pistols, and potentially even explosives.

The man carrying the suitcase walked into an apartment building and then came back out empty-handed.

The Libyans had just delivered weapons to West Berlin. Whatever their target was, they were now poised to attack it, and, with no reason to expect his superiors to intervene, Wiegand decided the only remaining move was to show the Libyans his cards.

He found Stuchly, convinced him that they needed to do something bold—something they'd need to keep especially covert—and that night they went out into East Berlin to an apartment used by an employee at the Libyan embassy whom Wiegand believed to actually be the Libyan spy agency's station chief.

They waited for him to come out, and when he did, they pounced. They shoved him back, pinned him in place, did their best to intimidate him, and, once satisfied he was sufficiently frightened, Wiegand told him he knew everything the Libyans were up to. He said other intelligence services knew as well and that if the Libyans carried out the attack he knew they were planning, those responsible would pay dearly. And then, as quick as they approached, they withdrew, leaving the Libyan station chief to consider their message. There was little to do now besides go home, hope they'd been persuasive, and to try to get some rest.

WIEGAND WOKE UP on the day of the planned attack without any new ideas for how to stop it. He drove to work, tuned a radio in his office to the West German police frequency, and with Stuchly alongside him, spent the day bracing for news of attack.

An hour went by with no emergencies reported by the West German police. Two hours passed; there was no news by noon. Still nothing by the evening.

By midnight there was still no radio activity concerning any kind of attack, and now it was no longer March 26. The date of the attack had come and gone, and West Berlin was still strangely quiet.

The surveillance reports Wiegand received the next day were even stranger: a woman with papers from the Libyan embassy had boarded the elevated train and headed toward the wall, crossed into West Berlin, proceeded directly to that same Kreuzberg neighborhood the men in the Volkswagen had gone to, walked into that same apartment, and walked out with the same suitcase full of weapons.

But rather than delivering it to a third party or bringing it to any of the West Berlin targets the Libyans had cased, she simply carried it back across the wall. Back into East Berlin.

The reports, at first, didn't make any sense. Why go through the trouble of sneaking a suitcase full of weapons across the wall to the West, only to bring it right back to the East? Some kind of dry run, just to make sure they could get the weapons across unsearched? Stuchly had no better explanation than Wiegand. They summoned their moles, who snuck out to meet them, and together the accounts of what the moles had witnessed over the previous several days confirmed that there had indeed been a plot and provided more detail about it: that Eter's team had been planning to attack a club called Stardust, but they'd called it off because there was too much heat from the authorities.

Wiegand, then, hadn't just identified the plot. He and Stuchly had stopped it.

There was finally something to celebrate, and a chance for relief from the weeks of constant hypervigilance. He called Stuchly into his office and popped a bottle of sparkling wine. It was perhaps a little anticlimactic. Their superiors were already beginning to castigate them for exaggerating a threat that never materialized, rather than congratulating them for having thwarted it. But that was to be expected. Wiegand figured it was worth toasting to saved lives, even if those lives were American. And even if the Stasi didn't appreciate it, he and his one trusted friend had prevented the Stasi from looking bad. The whole Communist Party maybe; the whole communist bloc—from looking bad.

He clinked glasses with Stuchly and drank, and it didn't occur to him until much later that the Libyan activity over the past two days may not have been a plot that was aborted but rather a diversion intended to occupy surveillance resources.

Because just then, via a separate route, Musbah Eter was carrying a detonator and time-delay device hidden in a pack of cigarettes into West Berlin. No one saw it.

7.

At twenty-six years old, Musbah Eter had sauntered into Berlin as a young agent with confidence and a cause, arriving like some young soldiers reaching the front lines of a righteous fight do: unable or unwilling to consider things like the morality of the tactics he'd been trained in. He'd swallowed the logic whole: his people were an oppressed minority, his country acting as a vanguard fighting for the weak, rising up to cast off the chains and throw them back in the lap of their oppressors.

That his mission was on behalf of Muammar al-Gaddafi—a leader who'd promoted himself to "colonel" and then spent the country's wealth on his own wardrobe, draping himself in gold-rimmed everything, tracksuits and flashy sunglasses, as much Elvis impersonator as statesman—well, it wasn't Eter's job to judge his sovereign's style choices. And as outrageous as Eter's leader looked and acted, Colonel Gaddafi had a grievance potent enough to pull people to the cause. Preaching about the West treating Arabs as inferiors, NATO bases on the African continent and their navies hovering off Africa's coasts—in Gaddafi's telling it was all obvious dominance, the chest-thumping of the larger primates trying to quiet the rest. That the West rallied behind Israel rather than Palestine was little more than the strong bullying the weak, and all of it called for a strong African nation to stand up for the downtrodden.

By the time Eter was dispatched to Berlin, his bosses believed they'd stumbled on a way to even the playing field. Libyan security services were notching small victories all over the world, and enemies didn't seem to have a way to respond. Eter's superiors believed they were succeeding in their solemn march toward freedom, emancipation, and dignity. The magnanimous colonel spoke of these things as if anointed, tapped by the Almighty to unite Arabs, unite Africans, draw power to the colonized, a rhinestoned Moses of the new afflicted class.

So here was Eter. A good soldier arriving to fight for something bigger than himself. But also not a fool.

Eter was a polymath. He was sophisticated. The son of a cattle breeder, clever enough to make it from his father's farmland in Libya first to Italy, where he'd studied electrical engineering, then into the field to see other countries as a journalist.

He'd fallen ill on a stint as a foreign correspondent in Chad in '84, but his government swooped in to save him, flying him not back to Libya but to the better medicine facilities in Germany. And even as a kid from the Libyan countryside raised around livestock, he found he got along just fine in European cities. He healed, he picked up some of the language, he picked up a woman.

Now, two years later, he was back in Germany, this time healthy, there to advance a cause and service his debt to the country that had saved him. He'd settled into his new cover as a diplomat at the Libyan embassy, which was really a beachhead for Libya's revolutionary countermarch against the West—Eter was told to call his office the Libyan People's Bureau. And as he settled into the world's capital city of espionage and began assembling a cell armed with a direct line to the chief of Libya's spy services, and with Gaddafi's plausible argument that the Libyans were an advance guard in a fight for the freedom of all weak peoples, he found recruitment wasn't so hard. Not just Berlin's Libyans but transplants from North Africa and from the Middle East, including one especially helpful Palestinian named Ali Chanaa, who agreed to join Eter's cell, and who had an apartment in West Berlin that Eter could use as a staging area.

When Eter snuck the detonator and timer in a cigarette pack across the wall to West Berlin, it was Ali Chanaa's apartment that served as the stash house.

Soon all the bomb components would be there, in one place, on Ali Chanaa's kitchen table. Which didn't seem to bother Chanaa. In fact, the Palestinian had proven to be an ideal co-conspirator. He'd shown sympathy to the cause and, perhaps even better, he was down on his luck. Chanaa was in the midst of a failing marriage to a German citizen named Verena, who was also Chanaa's partner in a restaurant that had now failed as well. They still lived together, failed spouses and failed business partners, but Eter had recognized an opportunity. Verena had been diagnosed with histrionic personality disorder and was moving in and out of depressive episodes, but Eter realized he could use all of it; he didn't see Verena as a problem.

Nor did he see that the same misfortune making his hosts susceptible to his influence had made them susceptible to others' as well. It didn't occur to Eter that Ali Chanaa might be playing both sides. That Ali Chanaa was also working with the Stasi, under the code name Alba.

8.

Paveway Assembly Operations
Texas Instruments
Sherman, Texas

With an optimized workforce largely comprising young women, and the cost of chips coming down, Paveway was proving cheap enough to seep into the different military branches.

Paveway was a weapon the military could justify purchasing in large enough numbers that they didn't need to be as stingy on practice ranges. Pilots could practice with it in live-fire exercises. While in Washington American military officials were trying to establish whether some kind of "preemptive" strike against Libya was possible, American pilots training at Nellis Air Force Base outside Las Vegas were going through so many Paveways that TI decided the company should have an employee stationed there.

Nellis Air Force Base had become a kind of finishing school for pilots practicing precision. Nellis was a squadron's last stop before deployment, a place pilots could drop live bombs in the closest thing to combat the Air Force could muster. Pilots practiced night missions; they learned to fly low enough to avoid radar; they ran bombing runs against heavily defended targets, with other planes pretending to be aggressor aircraft. And while they occasionally used some of the more expensive weapons TI's competitors were making, Paveway was their overwhelming preference. Air Force pilots dropped countless Paveways in practice, so many that by the time Reagan settled into his presidency—and by the time a stylish young agent named Musbah Eter landed in Berlin with a mission from his Libyan spymasters—Paveway had shifted from novelty to staple, taken for granted by a new generation of pilots for whom using the bomb was becoming almost second nature.

TI decided an employee out at Nellis could serve as a chaperone of sorts, for a weapon gaining independence, and customer support for a valued client. But Weldon saw another benefit. A TI consultant at the base could certainly answer technical questions for pilots, but could also be Weldon's eyes and ears, an intelligence asset embedded in a key community.

Soon, Weldon began to hear that pilots around Nellis were talking through ideas about new tactics. Paveway allowed pilots to stay at higher altitude, where it was harder for enemy defenses to reach them, but other countries were compensating by building better ground defenses. If other countries were getting better at shooting down planes at high altitude, pilots might have better luck doing the opposite: trying to stay low, below radar. The TI consultant reported that pilots were beginning to talk about the "pull-up" maneuver: coming in low, then pulling up at the last minute to "toss" the bomb upward so it would have enough altitude to reach its target.

Weldon considered the reports of conversations at Nellis he was getting, and saw another series of trends converging. He knew one of the big defense companies, General Dynamics, was building a new long-range plane called the F-111. He knew because Texas Instruments was building one of the components for the plane: a terrain-following radar system to help avoid obstacles, even when flying with poor visibility.

A plane equipped to avoid obstacles would be well suited to flying low, and therefore well suited for the kind of pull-up maneuver pilots were talking about.

Weldon could also see the military was so pleased with the success of Paveway that, despite his own doctrine of simplicity and affordability, he wouldn't be able to fend off the inevitable pressure for upgrades forever.

So he sat in his office pulling together threads, thinking it through, assembling trends: the need for an upgraded Paveway, the idea of a new tactic for releasing the bomb, a new plane ideally suited for just that tactic. Weldon decided he'd resisted long enough. It was time for Paveway to grow up. It was time for Paveway II.

WELDON SENT HIS team back to the "creative backlog" box with sheets full of sketched concepts they'd tossed aside during their sprint from prototype to production model. They chose the best ideas and attacked the engineering obstacles.

If Paveway was going to be released from low altitude, it would have less time to find its target. It needed to get better at flying, but Weldon couldn't make the fins bigger without taking up more space under the plane. One of the ideas in the backlog box provided a solution: retractable fins. With fins tucked in, Paveway II could actually take up *less* space, and cause less drag on the airplane, allowing for a modest but still meaningful improvement to the airplane's fuel efficiency, and therefore its range. Paveway's own range would improve because the fins that sprang out after the bomb was released would give it a wingspan almost double the original.

A new Paveway meant a chance for its brain to evolve too. The revolutionary silicon semiconductor chip had already allowed Texas Instruments to both drive a computing wave and ride it. The company was pushing the per-unit manufacturing cost of its chips lower by selling them in new weapons for the military and an expanding collection of products for civilians: the handheld calculator; wristwatches that replaced the movement and mechanical hands with semiconductor chips and digital screens; and, by the early eighties, the world's first 16-bit home computing device.

Others were joining the fray, as well: a new company founded to make new computing devices accessible to more people and given the accessible, unintimidating name Apple, had just released its second attempt at a mass-market computer, small and inexpensive enough to fit inside a home. A company called Atari followed suit; Tandy Radio Shack released one and Commodore started selling a version that gained a cult following, all built on the silicon semiconductor chip technology TI first sold to the military inside weapons like Paveway. And all of them using a promising new version of the technology that was beginning to mature at just the right time.

Thanks to one of Weldon's colleagues, a brilliant electrical engineer named Jack Kilby, Texas Instruments had a patent for something called the integrated circuit. If the first Texas Instruments chips let you use a small piece of electronics rather than a cumbersome vacuum tube to perform the function of stopping or amplifying a signal, the integrated circuit took it a step further. With even smaller, more *micro* components, the integrated circuit let you perform multiple such functions on one small, "micro" chip.

It had taken two decades to figure out how to make integrated circuit "microchips" effectively and inexpensively enough to be useful for more than just a few extravagantly funded military programs, but now they were accelerating a process of shrinking "computing" devices that would eventually make them small enough to fit not just into people's homes but ultimately into their hands. The integrated circuit would so thoroughly change the world that it would win a Nobel Prize in Physics, and Paveway II was among the first weapons to use it.

The newer integrated circuit microchip gave Paveway II's brain denser connections than Paveway I's. The evolved brain let it think more quickly, and it let Weldon teach Paveway new skills. He taught Paveway how to avoid distraction. While Paveway I followed any laser reflection, Paveway II could be set to follow only a laser reflection pulsing at a certain frequency. That could prevent an enemy from confusing Paveway or steering it off course by aiming their own laser. It also meant multiple Paveway IIs could attack different targets on the same mission, even if they were dropped at the same time, because they could each be set to follow different lasers pulsing at different frequencies.

With a maturing body and an updated brain allowing it to glide more efficiently, Paveway II could fly farther and was better suited for use at low altitude, which, in turn, reduced the time pilots needed to spend in dangerous airspace. Paveway II was even better suited for dangerous missions against well-defended, faraway targets, so when Pentagon officials tasked with confronting an emerging threat from Libya began trying to figure out how they might realistically launch some kind of preemptive strike against a dangerous target on the other side of the world, it was perhaps inevitable that Weldon and his engineers would start getting more and more calls with technical questions about just how capable this new model, Paveway II, really was.

9.

Building 2
Ministry of State Security HQ
East Berlin

Soon after Wiegand and Stuchly clinked glasses, celebrating the interrupted plot, Wiegand received a report that chatter between the Libyan embassy in Berlin and the Libyan spy agency's headquarters back in Libya was still heightened.

That wasn't the most encouraging news. A terror cell that had just aborted a mission should have been taking precautions or gone completely dark.

Worse, a report came in saying the Libyans were back out on the streets in West Berlin. It looked like they were casing nightclubs again.

Was the plot back on? Was Mustafa wrong?

Was it worse than that?

Wiegand now had to entertain the possibility that all the intel he'd gathered thus far had been a diversion. Could the Libyans be that clever? Could Mustafa, who'd been almost humorously easy to recruit—perhaps *too* easy to recruit—have been a dangle, part of a Libyan ploy to feed misinformation?

But Stuchly's mole had confirmed it all. Could Alba be a double agent too?

Wiegand called Stuchly back and castigated himself out loud for underestimating the Arab mind, which he now saw as even more devious than he'd once believed. How could he have been so easily duped?

Now he knew the stylish Libyan spy had the upper hand, and Wiegand was nearly out of options. But what was real and what had been fed to him to keep him off the scent?

If the plot was still underway and Wiegand was actually behind the eight ball, he'd need double, maybe triple the manpower to catch up and stop a Libyan threat, and however fantastical the chances of support from his higher-ups had been before, he knew there was no way he'd receive it now. He'd be lucky if they didn't throw him in jail.

It was time to break the glass and use the hidden communication line he'd set up as a contingency. One by one, he called in each of the six double agents he'd placed with the CIA and asked them to signal their handlers. He sent them over to West Berlin with all the intel he'd collected so far on the Libyan plot. His only hope of foiling the Arabs now was that somehow, by combining his information with whatever the CIA might already know, the Americans could figure out where the attack was going to happen and how to stop it.

10.

Residence of Ali Chanaa – Stasi code name "Alba"
Lindenstrasse, Kreuzberg
West Berlin

As CIA handlers in West Berlin began hearing from their assets about a plot against Americans, Musbah Eter's team at the Libyan embassy took receipt of 1,500 grams of plastic high explosives delivered via diplomatic courier.

Inside the small Kreuzberg apartment on Lindenstrasse, all the bomb components were now in one place. Eter gathered the group around the kitchen table and got to work. He took out a small block of the flexible plastic explosives and used a ballpoint pen to gouge out a small hole.

From the pack of cigarettes, he pulled out the detonator, a thin aluminum tube with exposed wires on one end, and inserted it into the hole. He took another device from the cigarette pack: a small, rectangular timer like a digital watch.

Hunched over the table, he wired the time-delay device to the detonator, just as he'd been trained. He pressed nails and small pieces of steel into the block. He wrapped the whole parcel in tape.

The bomb was now assembled.

Eter turned to Verena and addressed her in German. "You could help teach a lesson to the American guests," he said. Fate had called on her. Since she was a German citizen, she should be the one to carry the bomb. A German woman with a large bag would draw less attention than an Arab man would.

She seemed to consider it.

The men waited.

She pushed back. She had a better idea. Wouldn't two women be less conspicuous than one? She asked if her sister could join. Eter had no objection. He slipped the bomb into a travel bag, made sure Verena had taxi money, and told her that when it was done, she should not take a taxi directly home but instead follow an indirect route and switch taxis at least once.

Around 10:30 p.m., the two women walked out of the apartment into West Berlin.

Eter crossed back into East Berlin and his team sent a coded cable to tell Tripoli the plan was in motion.

11.

Berlin Operations Base
Central Intelligence Agency
Zehlendorf, West Berlin

CIA officers were already on high alert, having been informed by Wiegand's breathless double agents about a possible imminent attack, when they received a message about a telex British signals intelligence had intercepted and decoded:

tripoli will be happy when you see headlines tomorrow

The CIA passed it on to the U.S. military. The military sent an urgent warning to the Army's operations center in Germany, interrupting the commandant's dinner.

The commandant tried to narrow it down. The target could be anywhere and the attack could happen at any time—within minutes, for all he knew. He dispatched military police into the city, to every venue he could think of where American servicemen might be partying on a Friday night. As the commandant's men fanned out across West Berlin—running through the night from club to club, looking for anything suspicious, shooing Americans out and sending them home—the signals intelligence agencies intercepted another cable to Tripoli. Decoded and translated, it read:

happening now

12.

Friedenau District
West Berlin

Just before midnight, two women walked down a wide avenue in a middle-class neighborhood carrying a bomb the size of a milk carton hidden in a dark travel bag. They approached a six-story stucco building and entered under a red banner that read LA BELLE DISCO CLUB.

The La Belle discotheque was not an ideal target. It was frequented by Americans but also by Turks and Africans, even Arabs. The purer American targets were infeasible, though. Inside the U.S. military barracks, nightclub guests were nearly all American, but they had controlled-list entry. La Belle had an unfortunately diverse clientele, but, while casing potential targets, Eter had noticed that visitors there were given only cursory searches and sometimes none at all. He'd decided La Belle would have to do.

Once inside, Verena moved toward the bar, her sister at her elbow. The stools were mostly empty. They haggled over where to sit, decided on the corner of the bar, and Verena ordered a split of champagne. La Belle was one of the few bars in West Berlin open past 1 a.m., so the dance floor was still mostly empty. They waited.

The owner noticed two women sitting alone at the bar. He asked his bartender to make something special for them.

A man approached Verena's sister and asked her to dance. Verena stayed with the bomb.

People began to file in when other Berlin bars began to close. As La

Belle began to fill up, Verena kept her eyes on the bag. Soon more than a hundred people swirled around Verena, then two hundred, among them dozens of American soldiers.

By 1:30 a.m., the club was packed, and Verena decided it was time.

She signaled her sister. She reached down to the bag and set the timer. Together, they walked as calmly as they could out into the night, Verena flagged a taxi, and they pulled away from La Belle just as a team of MPs rounded a corner down the block and ran toward it. They'd been all over the city trying to find the bomb, but as the night went on, they'd come across more bars and restaurants starting to close and tried to narrow their search to places open late. One of them remembered La Belle.

Even from a distance the MPs could tell La Belle was crowded. It had to be the target.

They were just a few hundred yards from the entrance when glass blew out from the front of the club and scaffolding rods rocketed across the sidewalk. Inside, the DJ was thrown through the floor into the basement as the ceiling collapsed and the walls caved in. Hundreds of eardrums burst simultaneously as the lights went out, so the survivors felt they'd all suddenly been thrust underwater.

The MPs had been minutes too late.

13.

National Airborne Operations Center

Secretary of Defense Cap Weinberger was airborne, on his way to a meeting with Pacific allies, when he received a transmission about a bombing in West Berlin. He passed it to a deputy.

"Is this finally our smoking gun?" he asked.

The deputy read it. "I believe this *is* indeed a smoking gun."

By the time Weinberger made it back to Washington, the rest of the National Security Council had already gathered to study the Libyan intercepts.

There was little ambiguity. The Libyans had launched an attack deliberately targeting Americans, and Weinberger had no more argument for restraint. In the Situation Room, the president looked to his generals. It was time. Was there some way to launch an attack against a well-defended, distant target while avoiding extraordinary risk and excessive civilian casualties, all without much ground support or help from allies?

The generals had exactly the tool in mind.

They had a concept built around Paveway II: now better at gliding than Paveway I, so pilots could stay lower; improved eyesight, so they could pursue multiple targets at the same time and pilots could spend less time in dangerous airspace; and designed with a new plane in mind, capable of long range and equipped with a terrain-following radar so it could fly close to the ground even at night. The bulk of the airpower would come from the 48th Tactical Fighter Wing, whose pilots had practiced extensively with Paveway II during the TI-supervised training at Nellis Air Force Base outside Las Vegas. And the 48th was now stationed in Lakenheath, England, the one country they knew they could count on to cooperate.

PRESIDENT REAGAN APPROVED. But the mission had to be planned within the tightest circle of advisors. After all, pilots would be flying into enemy territory without any help from allies suppressing enemy air defenses. The element of surprise was critical not just to success but to their safety, and if the press got ahold of the story, they could doom the mission and cost pilots their lives. So even though this would be an act of war against a foreign state, President Reagan decided he couldn't risk informing his own Congress, not until the mission was well underway. He wanted it kept secret from much of the White House—for that matter, from national security advisors who didn't absolutely have to know. Even much of the Situation Room staff would be kept in the dark. He didn't need any of them. If the new Paveway was as simple, cheap, easy to use, and precise as the generals described, he could strike targets thousands of miles away with hardly any military mobilization. Paveway had enabled an exceptionally small number of people, huddling out of sight in an anonymous White House executive secretary's office, to draw up a major military operation on the other side of the world.

Within a few days the plan was ready. The operation, code-named "El Dorado Canyon," was put before President Reagan for final approval. Reagan gave the go order, and the first preemptive strike against a state sponsor of terror commenced. America went to war. Almost no one in America knew.

14.

Lakenheath Air Base, home of the 48th Tactical Fighter Wing
April 15, 1986
Ten days after the La Belle discotheque bombing

At 5:31 p.m. local time, planes began taking off from an airfield in England. Aerial refueling tankers, jets with radar-jamming equipment, and eighteen bombers: F-111 Aardvarks, carrying an arsenal consisting almost entirely of the latest Paveway II laser-guided bomb. The Aardvarks would have help from aerial refueling tankers and an aircraft carrier scrambling Navy jets to help suppress Libya's antiaircraft defenses, but otherwise they'd be alone.

Without cooperation from any country besides the United Kingdom and unable to use Italian, French, or Spanish airspace, the Air Force had to route the strike force out into the North Atlantic Ocean, around the Iberian Peninsula, then back through the Strait of Gibraltar, adding thousands of miles to the trip and requiring repeated midair refueling. The Aardvarks flew the entire trip with refueling tankers riding alongside them, continually connecting and disconnecting using the "mother tanker" concept that Rick Hilton had improvised over Vietnam fourteen years before. Just before the last Aardvark disconnected from the tanker for the last time outside Libyan airspace, jets from the aircraft carrier swooped in, jamming Libyan radar and dropping smaller payloads on Libyan antiaircraft positions before disappearing back out over the water. Then the Aardvarks split off and settled into their separate approaches.

One formation closed on a Libyan naval complex fifteen miles west of Tripoli, where Paveway IIs tumbled out of their weapon bays and followed lasers toward a fleet of training vessels and a series of outbuildings.

At the same time, six Aardvarks buzzed the Tripoli airport, bombing the runway and a formation of troop transport planes while three more flew toward the Tripoli barracks compound, the most difficult target since it was set inside a civilian area, but also the most important: one of Gaddafi's homes was inside the compound, as was the headquarters of the Libyan secret service. Paveway IIs struck various aimpoints in the compound, one after another, explosions marching right up and through Gaddafi's home and the spy agency headquarters.

The raid was frenetic, devastating, and overwhelming. The Libyan air defenses were caught outmatched. All but one of the Paveways scored a direct hit on its intended target, and by 2:13 a.m. local time—after only eleven minutes in Libyan airspace—all but one of the Aardvarks had made it safely to the rendezvous point. The planes banked back out around the Iberian Peninsula for the journey home.

15.

Oval Office
The White House

The minute the planes reached Libyan airspace, the need for secrecy disappeared. As the first reports filtered back to the Situation Room, excitement spread. Early intelligence analysis of infrared imagery seemed to reveal excellent effects around the key targets. A rumor passed through Washington that Gaddafi himself had been killed.

From the moment Americans began learning about the raid, support was nearly unanimous. Many Americans listened to it live as foreign correspondents locked in Tripoli hotels woke to roaring jets, called into their New York network headquarters, and held their phones out of windows. Americans who didn't catch the raid in real time soon saw a new national news anchor narrating it to them: the president himself, under the lights in the Oval Office in prime time, superseding all networks to deliver his good news.

"My fellow Americans," the president said. "At 7 o'clock this evening, Eastern time, air and naval forces of the United States launched a series of strikes against the headquarters, terrorist facilities, and military assets that support Muammar Gaddafi's subversive activities." The president was stern and edifying. "On April 5th, in West Berlin, a terrorist bomb exploded in a nightclub frequented by American servicemen. Sergeant Kenneth Ford and a young Turkish woman were killed, and 230 others were wounded, among them some 50 American military personnel. This monstrous brutality is but the latest act in Colonel Gaddafi's reign of terror."

The president looked into the camera and cocked his head like he was taking a child into confidence, privileging someone just old enough for these grave matters of American force.

And, now, he said, we had a new tool in our toolbox.

"We believe that this preemptive action against his terrorist installations will not only diminish Colonel Gaddafi's capacity to export terror, it will provide him with incentives and reasons to alter his criminal behavior . . . this mission, violent though it was, can bring closer a safer and more secure world for decent men and women." Preemptive strikes, he told viewers, really worked.

There was little question: there wasn't appetite for anything but triumph, and reports about the mission's shortcoming failed to gain traction. The loss of one plane and its crew was a negligible detail amid such a splashy debut for a new kind of war—as was the fact that at least thirty-seven people had been killed, or that one of the pilots had laser-designated the wrong aimpoint and nearly delivered a Paveway into the French embassy.

That people in Libya were already beginning, even before a full accounting of the destruction, to rally around their leader was a worry for another day; that when the body of an infant was discovered among the rubble, and state news announced that it was Colonel Gaddafi's own daughter—a child named Hana few knew existed, and some suspected had been adopted for propaganda purposes after her body was found—the leader's personal loss drew his people closer to him. Baby Hana became a rallying cry, and when the chief of the Libyan spy agency's operations division stood up before the screaming throngs at the mass funeral, he vowed that in Hana's name "we will take our revenge! We will kill Americans!" When he was done speaking, he came down from the dais and called his spies stationed around the world to begin planning terror attacks on an entirely new scale.

In America, the particular details from Libya could wait. Reagan was triumphant; his effective war and his speech explaining it set off a wave of patriotism. There weren't yet any pictures of what had happened on the ground; there was only Reagan. And "Reagan," the papers said, "was magnificent."

Preemptive strike was a shiny new silver bullet, and the president was a perfect salesman. "In a credibility contest, Reagan would put all three network anchormen out of business quicker than you can say A.C. Nielsen," the papers said. Congressional leaders lined up to praise the new tactic, showing themselves to be unbothered by the fact that a major military operation had been planned without them even knowing. "Failure to act after we had proof of Libyan terrorism would have given credence to the view that America lacks the will to stand up and fight against terrorism," Senator Alan Cranston, a longtime Reagan critic, told the *Los Angeles Times*. Others were nearly overheated with patriotic fervor. "I'm so proud that I'm an American, not like European countries that put economic interests above everything else," Congressman Robert Dornan said, calling the strike "biblical in its applications." Americans of every political stripe took notice. The success of the mission even began to inform a new generation of political hopefuls, among them a young governor in Little Rock, Arkansas, named William Jefferson Clinton, who followed the coverage closely and, impressed by what he saw, absorbed the idea that guided bombs might be just the tool to stop violent men.

Libyan People's Bureau
East Berlin

By the time Musbah Eter learned of the air raid, he was beginning to see cracks in his sovereign's logic. His chief, who had just pledged the deaths of hundreds of Americans in vengeance at the mass funeral, now passed orders on to Eter to start planning a grander attack.

Eter could see what was happening. Restraint was fraying. His nightclub bombing killed two people and left one fighting for his life, and many of the people injured were soldiers; the response had been an air raid that killed dozens of people and injured more than a hundred. Now his boss was telling him to double the number of casualties Libya had just suffered, and was beginning to plan an operation that would end up being the worst terror attack in the history of the United Kingdom: the bombing of a Pan Am flight headed to America that would kill 259 people aboard the plane and eleven on the ground in Lockerbie, Scotland.

There was only escalation now, no obvious off-ramp, so Eter began looking for one of his own.

He'd done his job; he had other things on his mind now. He decided to propose to his latest girlfriend, who was pregnant, and Eter had begun to think about a life back home in Libya. She was an East German citizen and not allowed to leave, but he had some experience smuggling contraband out of East Germany and came up with a wild idea. He hid her in a moving box, marked it as diplomatic baggage, and tried to sneak the giant parcel across the Berlin Wall to the West.

The ruse failed immediately. Border guards searched the box, found Eter's wife, arrested Eter, and extradited him back to Libya, where he was put in jail for seventeen days.

Had he waited just a little longer, things might have turned out differently. Demonstrations in the partitioned city were reaching a fever pitch, and a little while after Eter's failed attempt to smuggle his wife across the wall, images began circling the world of young people scrambling on top of it, sitting on it, kicking it, swinging at it with pickaxes and hammers. Everyone had wanted the Berlin Wall to come down, but when it finally fell, no one saw it coming.

Germany was opening back up, and once Eter was released from jail back in Libya, he returned to Germany to reunite with his wife. He tried again to bring his new family home to Libya; it worked for a short while, failing this time because his wife didn't care for Libya and wanted to stay in Berlin. For Eter, though, Berlin was becoming infeasible. It was only a matter of time before his role in the bombing was uncovered. His applications for travel documents to Germany kept getting rejected, and then, when Germany opened a new investigation of the La Belle bombing—thanks to new evidence provided by a disaffected Stasi colonel with a special interest in Eter himself—Eter learned a warrant for his arrest had been issued.

He came to see that the only course of action available to him was to collaborate. He walked into the German embassy in Malta, confessed to his role, and offered to cooperate with the investigation in the hope that it would allow him one day to live freely with his new family back in Germany.

WIEGAND, IN THE end, had made the same calculation Eter did. He'd made his own run at freedom as the Berlin Wall came down. As most Germans and much of the world sat in front of their television sets, watching the drama of the fall of the Berlin Wall play out, Wiegand started ferrying confidential documents out of Stasi headquarters. He loaded as many as he could into a small car, drove across the falling wall, and, once in West Germany, offered himself up as an informant. He requested immunity only for crimes he may have committed with the Stasi.

Wiegand was picked clean. West German intelligence debriefed him, then passed him to the CIA, who passed him to British intelligence, who handed him off to French intelligence, all of them finding him to be a model defector who provided, among the wealth of stolen documents, a mountain of damning evidence against the Libyans—and against one young agent in particular, Musbah Abdulgasem Eter—for an attack Wiegand had spent the end of his career trying to stop.

Together, Lieutenant Colonel Rainer Wiegand of the East German Ministry of State Security and Musbah Abdulgasem Eter of the Libyan secret service operations division, helped investigators bring the La Belle bombing case to trial. And it would be at the trial that the two spies who'd operated in each other's shadow for years might finally meet face-to-face. Eter was slated to testify about his agency's acts of terrorism, Wiegand to testify that his agency knowingly let those attacks happen. Both were taking risks, Wiegand perhaps a more significant one, since he would be implicating one of the biggest, most feared, and most powerful intelligence services ever created.

So it was perhaps unsurprising that Wiegand never made it to court. Just before he was due to testify, he was killed in a car crash under mysterious circumstances.

16.

Defense Systems & Electronics Group
Texas Instruments
Lewisville, Texas

The day after the raid on Libya, outside Weldon Word's office, a crowd gathered in the visitors' lobby. Early reports held that Paveway II had worked exactly as it was supposed to. The Air Force would surely show up in a week or so for a technical debrief, and they'd have video from the targeting pods, but no one wanted to wait that long. A junior staffer ran out to one of the local network affiliates to beg the daytime producer for tape of the overnight network feed, and soon a gaggle of TI executives and engineers were gathered in a conference room, watching footage from the Aardvark targeting pods. Black-and-white structures, anonymous factories of terror, turned in a flash to ash and cloud. Paveway, a thing they'd all touched—a thing Weldon had dreamed into being—out there, doing that. Quieting a madman, shaping the world.

For Weldon and his team, it was proof that a version of the future they'd foreseen was arriving. TI's history had been rewritten. They weren't oil prospectors anymore; this wasn't the 1940s company rendered obsolete by a future war. This was a company that *enabled* future war.

RICK HILTON, NOW a decade and a half from presiding over Paveway's coming-out party over the Dragon's Jaw, began hearing details about its latest mission from his community of retired pilots.

He wasn't happy, though: something was a little off. Rather than pride, he felt something closer to foreboding. It was good that skepticism about precision bombing was fading, but what arose in its place wasn't what he'd hoped for.

He heard friends talking about how one of the F-111s, call sign Karma 51, hadn't come back. Hilton knew all about the Aardvark. He knew about the new terrain-following radar. He knew the plane could fly low. He could close his eyes and almost imagine what happened to that one plane. He didn't know any of the pilots personally, he didn't judge them. This wasn't his war. But he could see a lesson had been learned since he'd dropped the Dragon's Jaw, and he didn't think it was the right one.

He knew what happened to that one plane. He was as certain as he could be without actually having the data in front of him. Karma 51 had crashed while trying to avoid enemy defenses. Those pilots had died trying to stay safe. He could picture it: pilots trying so hard to eliminate risk that they eliminated it right into the water.

Hilton saw where war was going. A lesson taught by Paveway, which was beginning to convince people they could fight a war without really fighting a war. Hilton didn't believe war could be done cleanly. Violence was messy; to try to make it otherwise was folly. Since that one final mission in '72, dropping those first bombs, deciding to stay in a dive and fly down into range of all the ground defenses so Bill in the back seat could keep the laser light shining on the target—you ignore risk for the moment because a warrior should be dedicated to the mission first, his own survival important but secondary. Problems arose when priorities wobbled out of balance. He thought this business of trying to make war safe and clean was a distortion of the natural order of things, and sometimes got you right into the state you were trying to avoid anyway, which was dead.

If what he guessed was right, what he saw wasn't just bad tactics. What he saw was a weapon beginning to bend war planners to its will.

BOOK IV

KATHY

(OPERATION DESERT STORM)

1.

Lexington Federal Prison
Lexington, Kentucky

A woman sits in a prison.

She is a radical. She can preach her beliefs. She's studied the faith; she's fluent in it. She'll soon decide she is willing to become a martyr. She is willing to go to war.

Then, soon enough, she will.

IN PRISON HER name, Kathy Kelly, is taken away and replaced by a number. Then replaced again by a nickname. Other inmates can't seem to believe the reason she's here. They can't get their minds around it. She explains, they shake their heads, she swears: *Yes, I did, I broke into a nuclear missile silo. Over in Missouri, I planted five kernels of corn.* They shake their heads some more, like they couldn't possibly have heard right. What might possess someone to do that?

Hey, Missiles, they say, *tell us the story*, and they laugh at her, but at least she's not anonymous. *OK, Missiles,* they say, *tell us the story again.*

SO SHE DOES.

A protest movement she'd practiced for; her and thirteen others. She spent the whole summer of '88 bicycling around her hometown, looking for abandoned industrial lots to practice breaking and entering.

Mostly it was the barbed wire that frightened her. Would she have to vault herself over a fence? Just over five feet and a hundred pounds soaking wet, a little less if you didn't count her most imposing feature, a shock of wild hair. What if she got stuck or fell and broke a bone? But it was that or breach the perimeter fence with a lock-breaking tool she wasn't sure she was strong enough to use.

So the day comes: a mid-August Missouri scorcher, that moist hugging heat. Her heart hammers in her chest as she's driven to the site, because she still doesn't know how she's getting in.

They reach the perimeter, and her driver, a friend she didn't want involved beyond the drop-off, decides he'll help. He hops out, grabs the two-handled tool, breaks the lock, and takes off. Kathy walks onto Whiteman Air Force Base.

Inside the perimeter, she moves through a meadow of shin-high weeds, over gravel, and to a concrete pad, just a lip above the earth.

Mist rises around her; maybe some kind of coolant.

She knows, vaguely, what's under her. Down there in the dark, just out of sight, a nuclear-tipped missile points up, bigger across than she is tall, and almost sixty feet from bottom to top. A launch order from the president and the pad slides open and the earth squeezes the missile out, a 70,000-pound excretion already miles in the sky in the time it takes to blink. Crashing into the exosphere at twenty times the speed of sound, God-like power. Faster than the human eye can see.

She doesn't yet know that this missile reaching eight stories below is driven by technology she'll encounter later in smaller bombs, steered by currents twinkling across its body in chips designed by a band of risk-takers looking for oil in Texas.

That this weapon, the Minuteman intercontinental ballistic missile, is built off the blueprint for the vengeance weapons Hitler used to terrorize London during the Blitz, including her own still-unacquainted parents, running underground to hide and finding each other there. She knows this missile is the reason she's here. She doesn't yet know it's the reason she's *here.*

She sits on the concrete and begins the next part of the plan. She unzips her backpack and pulls out a banner.

YOU CAN'T HUG A CHILD WITH NUCLEAR ARMS.

She takes out a handful of World War II medals a veteran handed her at some point before in the plotting: *I hereby renounce my support for U.S. nuclear weapons policy.*

She takes out a tablespoon and the most important thing: five kernels of corn.

She rakes some dirt aside with her fingers and uses the spoon to press each kernel into the ground.

And then she just sits.

For a while, it's quiet. Has no one seen her?

The mist rises, seeping through cracks in the ground, the exhale of a resting giant. She becomes aware of crickets. Cicadas? Some kind of clicking insect, heat activated. She's aware of the smell of grass, and birdsong. Have they grown louder, or has she tilted into some altered state? She sits on the ground, ten feet above an instrument capable of eighty Hiroshimas, and nature seems to be screaming at her.

A growl in the distance, this time a vehicle, a military truck approaching, all straining engine and moving aluminum. The truck hurries to her, pulling spirals of kicked-up dust, a machine gun mounted on the top. Three soldiers in full combat gear leap out and surround her like they're invading hostile territory. Crouching down, weapons raised. Arching their necks and jabbing at radios. She catches bits of transmissions.

"All personnel"

". . . clear the site"

They find no invading army, no foreign spies, just her. A tiny thin wisp of a woman, five feet and flare of hair, sitting on the pad like a schoolkid waiting for a late parent.

"Raise your arms"

"Step to the left"

They order her toward them. One of the soldiers has handcuffs. He tells her to kneel.

Another pause. Soon she hears the engine rev and crunching gravel behind her. The vehicle is leaving, but she doesn't know where they're going or why they've left her. It's entirely quiet now; even the animals have stopped making noise. But she senses she isn't alone. She feels someone has perhaps been left behind, the presence of an electricity that's not her own.

She senses a gun barrel at her spine. She doesn't move.

Now, in the fuller silence, she's feeling more uncertain. This exciting activism of hers has become real. Now she's in the heated part of the dream where you've stepped off a ledge, gone into freefall, and there's no turning back, all there is to do is await punishment.

Maybe speaking will interrupt worry. She speaks. To no one, to the clicking insects, to the soldier she thinks is standing behind her. Might as well start practicing her defense. She doesn't know what he knows, what versions of the story his training may have fed him. The power of these things, the scale of the project they're part of, the truer purpose. The thousand silos marching across the Great Plains, holding missiles intended not just to strike the Russians but there in the hopes the Russians strike *them*. Someone far up the chain of command having decided that if there's war with Russia, the New Yorks and San Franciscos have to be saved. Having decided the culture has to endure but that the middle states are expendable. A thousand nuclear missiles in the heartland so the enemy will have to hit the heartland first. The man behind her not just protecting a weapon that can kill people far away. A man who's here to maintain the bait.

She recites the statistics she's memorized. The cost, and what those dollars could've done in her own neighborhood. What if all this money, all this power, had gone into gardens, schools, libraries, into calculators for students? How many people could be fed if instead of solid fuel boosters and fissionable material that money was spent on food?

The soldier breathes behind her; the missile does something like breathing beneath her. Did he shift back there? She's made her case, more or less. She stops talking. She wonders if he's been listening at all. She can somehow tell he's young.

"So." She gropes for something else to say. She looks down to the dirt, where she's raked her fingers for little seedbeds. "Do you think the corn will grow?"

He doesn't answer at first. But after a pause she feels a shift. And now, for the first time, she hears his voice. He sounds younger even than she'd have guessed, probably not so long out of high school.

"I don't know, ma'am," the soldier says, and it's quiet again.

Then, after a time: "I don't know, but I sure hope so."

SHE REPORTED TO county lockup in bad shape. After the trial, before serving her sentence, she suffered a partial lung collapse and needed an operation to clear diseased tissue. She emerged from surgery with pink eye, a respiratory infection, and an unsettling watery cough.

At intake a guard relieved her of her personal effects, handed her a prison-issue mattress casing, then pushed her along into the "bullpen," a semipermanent home for a dozen women all bunkered together, an exposed toilet and six bunk beds stacked on not enough floor space.

She looked up at the one vacant bed, a top bunk that might as well have been the summit of Everest. Her eye itched, her arm throbbed, she tried and failed to stifle a cough, all heads rolled toward her, and someone yelled, "I don't know what I did so wrong to be locked up with this white motherfucker that got AIDS."

She fought tears; she felt like prey. She managed, "I don't have AIDS!" And that was enough to tip her into a coughing fit.

When she got her first phone time, it was Karl she called.

Karl, who burned white-hot in her life for long enough to illuminate a future version of her, the warehoused mannequin lit up from the back shadows. Karl's quiet kilowatt glow, holding just long enough to show her a goal and then leaving her to find her own way toward it, an amicable separation and his work here done.

Karl: an enigma of a man, anarchist/activist/roller skater. Gliding into meetings like a trick of the light. She'd met him back when she was cleaning urinals at the soup kitchen. That phase of hers when she leaned close because *God hears the cries of the poor* and she felt she should too. Bent over, bathed in the ammonia odor of mission work. In he'd rolled, a legend in shabby plaid, asking in his almost silent way, like speaking to his own chest, if anyone wanted to join him in getting arrested for some higher cause or another. She'd pinched off her rubber gloves and turned a page to a new chapter of life.

In jail, on the phone, she tried to keep from crying. "I'm not sure I can do this."

She heard the ruffle of the receiver moving around. Karl, quiet and matter-of-fact. "You've done things you didn't want to do before. When you had a choice. So I don't see why you can't do this now."

She waited for the reassuring part.

"When you don't have a choice."

That was all he had for her. So few words from a friend, a comrade, from a lover, but they steadied her; they stuck with her. After she hung up, as she limped back to the bullpen and tried not to tripwire a brawl with the bigger women. *When you don't have a choice.*

She was kept alongside people treated like debris, drifting into criminality because there was nowhere else for them to go. Society provided no outflow for people like these. You could make six figures as a contractor for the military, an engineer, building things to end countries. Six figures and the prestige of your peers, while her cellmates were sealed away from their children because they got caught with a plant on some agency's schedule of forbidden things. Which of these inmates posed a threat? To whom, exactly? If you built a nuclear bomb, *then* how many people do you threaten? If you started a war? Those people had a choice.

And it was once she understood this idea—*choice*—that she had none now, that she'd had it before, that the people she was with seemed to have maybe never really had one at all, that she started to feel she could make peace in the bullpen, make friends in jail, that she understood, that she knew there'd always be limits to her understanding.

She made a point of sharing the "week at a glance" planner she'd been allowed to bring with her. It had a small map as an insert, and the women took turns tracing their fingers along the paper to see where they'd each be going next. They began asking her personal questions. She was so unlike almost everyone else. Sickly, tiny, a teacher. Most couldn't believe she was there for planting corn over a missile silo. It amused them. They berated her. Who had the time to protest missiles? What kind of luxury was that? Who in their right mind? When the "Missiles" nickname began to catch on, it seemed she'd been accepted. They maybe even liked her. "I tried my hardest not to like you, Missiles"—Lorraine, who'd terrified her that first day—"but I just can't help myself: I like you."

If they were leaning in, so would she. The women in the bullpen never had enough utensils to eat with, toilet paper, no cleaning supplies, so she took action. When women came in on briefer stints, she used her diminutive size and her brightest smile to make requests. She asked them to leave behind things they might not need but the longer-term bullpen residents could use. She'd ask them to leave newspapers so the women could use them for toilet paper, until the prison major found out and lined the women up. "Which one of you all bitches had the nerve to say we don't give you toilet paper?" A moment for these new allies to hold the line, show resolve, for each woman to say, "I did it."

"Musta been Missiles," one of them said. "She thinks she's living in some kind of hotel."

This burned. It burned more maybe because it was probably true. She'd known she wouldn't be here forever; not everyone else did. They didn't all know if they'd ever get out, as she knew she would. She was probably the only one who'd come here more or less by choice. The others, some of them, maybe most of them, had committed crimes because crime seemed like the only way to feed themselves or their children.

That's what stuck with her, for the rest of her stay at county lockup. It stuck with her during her transfer to the maximum-security prison in Kentucky where she spent the rest of her sentence, and when it came time for her release, it was this that wrecked her. On her last day, on work assignment in the prison laundry, Kathy fell apart. Hiding sobs behind giant tumbling machines. She'd carry a guilt with her, for all the people she left behind. For all the people who hadn't really had a choice.

And it stuck with her when, soon after she was released—sentence served, "Missiles" nickname affixed—she began to see footage on the news of tanks rolling across a desert a world away. A dictator named Saddam invading a nation called Kuwait. In America, all around her the drums of war came to life. But Kathy had little trust in those who controlled the levers of power.

She began to read, and the picture that emerged was a little different from the one it seemed all of America was seeing. She saw two countries whose relations had been deteriorating for years. Saddam had assembled a roster of grievances, saying Kuwait manipulated oil prices, accusing Kuwait of "slant" drilling—sliding rigs under the border to suck up mineral resources that belonged to Iraq. Kuwait refused to forgive a loan Iraq had used to defend both countries against Iran, their shared rival. And Saddam said that were it not for British colonial interference, Kuwait would have been part of Iraq anyway, so to him sending tanks there wasn't expansion. It was restoration.

Kathy watched the beautiful Kuwaiti girl who appeared on the news, *Nayirah*, even her name was pretty, telling of heartbreaking things happening to the Kuwaiti people under Saddam's invasion. In front of Congress, she spoke of Iraqi soldiers raiding Kuwaiti hospitals and stealing incubators; how they'd thrown babies on the floor. It seemed extravagantly cruel. Nayirah—no last name given, testimony not under oath—wiped tears from her cheeks, her story so terrible, her delivery so compelling, that any American watching needed no more convincing to sentence a monster like Saddam Hussein to the kind of precise and righteous violence America now knew, after its triumphant operation in Libya, it could deliver.

But war never served the people you thought it did, Kathy thought, even when it seemed just. So, as the American public leaned toward punishment for another madman, she began to spin an idea around in her mind.

What was it about choice? She thought of the polite young soldier at the missile silo. *I don't know, ma'am . . . but I sure hope so.*

She didn't know his story, but he probably didn't feel he had much of a choice to be there that day—certainly not as much of one as she had. She thought of the phone call with Karl, and now she thought of something a friend of his had said—maybe where he'd gotten the idea from in the first place. Father Daniel Berrigan, the famous Jesuit priest. "Soldiers don't really have a choice; they get sent off to war and they have to risk their lives."

Soldiers didn't get to say *It's unfair to my family* or *It's important I finish my education first* or even *It sounds too dangerous; this isn't worth risking my life for.* Maybe as long as protestors felt it was up to them, and soldiers knew it wasn't, wars would keep getting fought and weapons would keep getting built.

But what if people trying to stop wars had the same mindset as the people fighting them? What if even the risk of death was not enough to give you a choice in the matter?

She watched Nayirah; she read about ant farm bases where American soldiers were already packing cargo planes and getting ready to deploy. Maybe, she thought, this was the way to finally change things. If the soldiers were already leaving for the desert—well, maybe a team of peace activists should go too. A team of assembled veterans from other protests. They would be a SEAL team without guns, radicals with go bags. *We accept that we may pay the ultimate price,* they took up as their unofficial oath. *We accept that we may pay the price that's demanded of soldiers.*

A war was forming. If soldiers were leaving for Iraq to fight—well, fuck it. A team of activists would go there to stop them.

2.

"Checkmate" Division
The Pentagon
Washington, D.C.

Right around the time Kathy Kelly was finishing her prison sentence and beginning to form a new idea for how to get in the military's way, a bookish colonel named John Warden walked into a small room in the basement of the Pentagon, flicked on the lights, and looked around. He surveyed the furniture and decided this would do just fine as his base of operations. This would be where he'd activate a secret planning cell, code name "CHECKMATE," which had only ever been used before for war games. This time would be different. This time he'd use CHECKMATE to plan an actual war.

John Warden believed the world had reached a turning point.

He'd studied airpower his entire professional life, ever since his days flying missions over Vietnam with pilots like Rick Hilton. Like Hilton, his worldview was shaped by witnessing the growth of precision bombing from its earliest days. Also like Hilton, it was shaped by seeing how the hesitant men in Washington tended to make everything worse, for everyone.

Now, Colonel John Warden was reaching the end of a thirty-year career, and here was a chance for his swan song to be the ushering in of a new, more correct era. He'd achieved renown as an expert on airpower, he'd written books about it, and when the U.S. military launched its air raid against Libya three years before, what he'd seen was a glimpse of the future. It just needed a nudge. And here, as his staff began arranging the furniture in the secret basement room, he was in a position to apply that nudge. He had the platform, as head of the Air Force Directorate of Warfighting Concepts. He had the opening, in Saddam Hussein's invasion of Kuwait.

The way Warden saw it: if Libya had been a preemptive strike to punish a dictator and prevent more terrorism, this was a chance to go one step further. Warden believed precision air strikes were capable of more than just punishing a faraway leader. He thought they could eliminate one. As the CHECKMATE team got to work, that was his unspoken goal: decapitation.

Warden wouldn't say it publicly, and when compelled to discuss the possibility of removing Saddam Hussein from power, he'd just say that the "decapitation" of Iraq was not a formal objective, but their plan just might yield it as a nice by-product. And he was keenly aware that decapitation had been a dream of war planners for perhaps as long as war existed—one rarely, perhaps never, achievable. After all, few had the access or the physical tools to kill a rival leader without a full-scale invasion first.

Now Warden thought he did.

Like the Operation El Dorado Canyon raid in Libya, his concept for Iraq required no ground troops. He'd minimize the risk to Iraqi civilians, too, because while the war would be devastating, it would be precise and it would be fast. Colonel Warden wanted to minimize collateral damage, but he would not go soft.

In his view, restrictive rules of engagement from overcautious politicians had prolonged the Vietnam War, cost more American lives, and in the end probably cost more Vietnamese lives too. The gloves would come off, and the war would be more humane for it. Warden would not have American pilots dying because the enemy had a chance to get its defenses back up and running. This war had to degrade Saddam's war-making capability as quickly as possible.

His idea was a campaign that was precise, but massive; devastating, but nearly instantaneous. Entirely from the air, like thunder. They would call it Operation Instant Thunder.

Once Colonel Warden's Checkmate team had a rough draft of a war plan for Operation Instant Thunder, he ran it by the CENTCOM commander, General Norman Schwarzkopf Jr., who liked the idea of an air-only war. The chairman of the Joint Chiefs of Staff, General Colin Powell, liked it, too, but hesitated; he argued some ground troops were still necessary, if only to confirm the air campaign had been successful. Colonel Warden adjusted the plan, adding cruise missiles from the Navy and a Marine contingent on the ground, turning an air-only campaign into an air-mostly campaign. No longer only thunder, still mostly a storm, but "Instant Storm" didn't sound quite right. So they replaced the descriptor of the tactic with a descriptor of the theater, and Operation Instant Thunder became Operation Desert Storm.

AFTER THE LIBYA raids three years earlier, there was pride and delight at TI headquarters in Texas, and at the company's first foreign factory in the sleepy little knoll of Bedfordshire, England, where workers left their assembly-line posts one day and went home to watch news reports of Paveways falling on military barracks in an exotic country, a madman terrorist quieted by something they'd sculpted with their own fingers. The new weapon was continuing to prove itself better in almost every way. Gliding farther, thinking faster, enabling new kinds of strikes, and keeping pilots even safer in the process.

Weldon Word, now a Texas Instruments vice president and a plant manager, vaulted by his offspring's success through a series of promotions, still thought Paveway could improve. The United States had lost only one plane on the Libya mission, but that was still two pilots dead, two families grieving. And while almost all the bombs had hit their targets, some had flown wide. Paveway II was better but not as good as it could be.

He still thought Paveway's chief advantage was its cost and simplicity but felt he couldn't be complacent. When you used an effective weapon, enemies adjusted to defend against it. When you improved it, they adjusted again. While Texas Instruments was shipping Paveway II to customers around the world, providing American allies with the ability to bomb targets from low altitude, Russia was upgrading its antiaircraft cannons and distributing them all over the Eastern Bloc. They built better short-range surface-to-air missiles to challenge new low-flying aircraft. They built automatic tracking and aiming into their antiaircraft weapons and developed ways to track incoming planes without a pilot's radar-detecting instruments registering the signal. Eastern Bloc countries were also beginning to use a more mobile surface-to-air missile platform that was easier to hide.

Bombing had become dangerous again, even for pilots dropping Paveways. Even when avoiding detection by flying low, pilots still had to pull their planes up to toss their bombs, gaining altitude and losing airspeed. New enemy methods made the "pull-up" maneuver more dangerous, but, without an upward toss, Paveway couldn't glide to a target that was any meaningful distance away. And that's where Weldon saw an opportunity. He could picture pilots at some point soon beginning to ask, *Why do I have to pull up at all? Why can't I just stay* low?

So how might a new Paveway, Paveway III, enable pilots to not just approach a target at low altitude but *stay low* while dropping their bombs? Weldon challenged his team to work the problem. They could try giving Paveway powered flight—a jet engine—but that would make it into an entirely different weapon. Certainly far more expensive, certainly more complicated to maintain, and potentially not as foolproof. To Weldon's team, that really just left one possible solution: Paveway would have to learn how to "pull up" on its own.

The challenges were immense. They'd need to teach Paveway III how to count, because it would have to wait before pulling up, or it'd risk pulling right back up into the underbelly of the plane. But every second it waited was more lost airspeed and altitude, so Paveway III would somehow need to be even better at gliding, even though its body had already filled out in all the ways it could to maximize aerodynamics. They couldn't give it bigger fins, at least not without requiring changes to the planes carrying it. Paveway needed to perform better but with the same body.

So they focused on how it *used* that body. Paveway I and II flew through the air with "bang-bang" steering—sending fins all the way in one direction or all the way in the other, like a first-time driver yanking the steering wheel too hard to the right, then too hard to the left. The team figured they could use the company's still-shrinking silicon microchips to help Paveway become more coordinated. They could teach it smoother, more precise steering. If they could get Paveway III to move its fins just a little at a time—to make subtler adjustments, smaller and more proportional to the course correction required—the zigzag flight pattern to the target would smooth out into a straighter line, allowing for a farther flight.

That still left the most important problem of all. For Paveway III to steer up on its own, it would need a new kind of self-awareness. All Paveway had ever needed to know was where the laser reflection was. If Paveway III was going to pull up, it needed to know something it never had to know before: which way was up. For that, Paveway would need an expensive education. Weldon needed someone to pay for it.

ON THIS LATEST project, at least, the Air Force turned out to be a less generous benefactor than Weldon hoped. He had a track record for anticipating future needs, but here he found himself looking perhaps a little too far into the future. He was proposing a pop-up capability that pilots hadn't actually asked for yet, and the Air Force wavered. So Weldon came up with another way.

He called a meeting of TI execs to make a pitch, and once they were gathered in the TI conference room, he opened with a recap.

"Paveway is a pretty nice-size piece of business," he said, leading with obvious understatement: everyone in the room knew Weldon's invention had revolutionized warfare twice, given Texas Instruments a near monopoly on precision strike, and helped bring the cost of microchips low enough to spark a personal-computing revolution. It was in part thanks to Weldon that the company was making money on calculators, computers, and an expanding number of products in entirely new sectors. He was the TI golden boy, or at least Paveway was, and Weldon was its handler. "I want to take some of the profits from that," he said, "and use it for company-funded research and development."

The Texas Instruments executives liked the idea of a true low-level release that might one day save pilots' lives, but they too were skittish about spending money on a product no one had asked for and that its most obvious customer wasn't willing to invest much in. It didn't seem like the most sensible business decision, especially when Paveway I and II were still selling well.

But Weldon had earned the right to make a bet or two, so they decided to offer a compromise. They'd do him the courtesy of loosening the purse strings, as much a gesture as anything else. It wasn't hard, from Weldon's position, to see the bet *they* were making: that if they indulged him a little—if they gave him, in essence, a sandbox to play around in with his newest passion project but not anywhere near the funding he would need to see the "pop-up" Paveway through to completion—perhaps he'd eventually just lose interest on his own.

He didn't. And almost immediately Weldon's latest reading of the warrior's aura started to look like a money pit. His team made progress too slowly, there was still no guarantee they'd be able to get Paveway to pull up, and even if it did, no guarantee the Air Force or any other branch would actually want it. Management at Texas Instruments grew nervous. Weldon showed no sign of tiring but also no sign of a major breakthrough. There didn't seem to be any trademark Weldon Word miracles in the works. TI execs decided they had to shut it down, but they needed a diplomatic way to do it. So rather than delivering the news themselves, they called in auditors to review the new Paveway program. A team would come in, crunch numbers, survey markets, and report back on how poor of an investment Weldon's new venture was, how unfavorable its financial outlook, and the bosses could pretend to regret the fact that they had to close the books on it.

But when the auditors reported their findings, it didn't go as expected. Instead of recommending TI stop wasting money on Weldon's vision, they recommended an *increase* in funding. Some of the auditors were so enthusiastic, they asked to be reassigned to Weldon's team and, when pressed, said it wasn't just the promise of Weldon's idea or his team's progress or even the value of a bomb pulling up on its own. What most excited the auditors was something entirely outside the company—something not even Weldon had paid all that much attention to but that, once he did, would lead to perhaps the most consequential new capability for Paveway since Rick Hilton's mission over the Dragon's Jaw.

DURING THEIR RESEARCH, the auditors had learned about an at-first-glance-unrelated capability the military was interested in.

The Air Force had been running a secret, special-access black program trying to develop a different way to deal with improving antiaircraft defenses. They'd contracted with the secret Skunk Works division inside Lockheed Martin to develop a plane that, using futuristic stealth technology, would be invisible to radar.

Pilots flying a true stealth plane invisible to radar wouldn't have to worry about staying low, which meant that, if the program was successful, the Air Force would be able to address one of the other ways adversaries had begun compensating for better American bombs: with stronger bunkers. Countries were bringing in specialists to harden bomb shelters and build better new ones, and while the few officials who knew about the stealth plane program spoke about it with almost no one, they were less restricted about their interest in a weapon that might service increasingly hardened targets. They wanted a bomb that could penetrate even reinforced concrete, survive impact, and explode only once it had burrowed deep enough to destroy a structure from the inside. The TI auditors had gleaned enough about the weapon the Air Force was looking for to anticipate one of the key challenges likely to stand in the way. Even if someone figured out how to build an aerodynamic shell strong enough to penetrate hard surfaces, that might solve only part of the problem. Getting a bomb to penetrate a reinforced surface would follow a logic similar to hammering a nail: no matter how strong and sharp the nail is, it won't penetrate wood if simply tossed at it. No matter how sturdy a bomb was, it wouldn't penetrate a reinforced surface if it didn't strike with its body positioned just so, angled directly at the target, approaching with its momentum driving as close to directly *down* as possible. And for that, a bomb would need to know which way was down.

It would need, in other words, exactly the capability Weldon was trying to develop for an entirely different purpose. The auditors pushing TI to increase Weldon's funding saw the chance for a marriage between the knowledge Weldon was trying to give his bomb and a capability the military wanted.

Shortly after the auditors delivered their report and gave Weldon's latest project a new lease on life, TI started to hear whispers that a penetrator warhead was indeed under development. One that used the latest materials-science research to produce metals capable of surviving impact with hard surfaces, but that, sure enough, still required a capability like the one Weldon's team was working on in order to actually penetrate.

Without realizing it, Weldon's team had been designing the technology for a weapon capable of defeating the very bunkers built to defend against it. Soon the connection was made, the Air Force read a small number of Paveway team members into the secret stealth plane program, and when the plane became operational, so did a special new "Paveway III," and with it a new class of bomb: the "bunker buster."

3.

Checkmate Division
The Pentagon

Colonel John Warden began to staff up. To help with the specifics of modifying a one-night attack into a multiweek operation, he brought in two of the pilots who'd flown in El Dorado Canyon against Libya and one of the officers who'd helped plan it.

He sent a contingent of CHECKMATE staff to work out of Central Command in Riyadh, Saudi Arabia, closer to the theater, where they made space for a vault in a third-floor office, gave it an anodyne name to avoid suspicion, and kept their work at the new "Special Studies Division" so secret, even from other CENTCOM staff, that outsiders took to calling the basement planning room the "Black Hole."

The two cells established a protocol. CHECKMATE, under the Pentagon, came up with target categories: barracks, telecommunication hubs, intelligence nodes. Black Hole staff under CENTCOM in Saudi Arabia took those categories and looked for specific buildings in Iraq. They accumulated sources, consulted spies, and tracked down defectors. They went out and interviewed builders who had worked on Baghdad's telecommunication network as well as engineers who'd built part of the city's infrastructure. They began adding potential target buildings to a list on the wall.

As planning progressed, John Warden remained insistent that Iraq's war-making ability be destroyed and that no hesitation or half measures allowed Saddam to threaten American pilots. Warden wanted to be decisive about destroying Iraq's military and intelligence centers but also the secondary facilities they might move to if the primary ones were destroyed. Hitler had his *Führerbunker* under Berlin after all, Stalin his air raid shelter near the Kazakhstan border. Destroying Saddam's command-and-control capability wouldn't cripple him if he could simply move elsewhere and wage war from there.

Finding backup facilities was more difficult. It wasn't just intelligence gathering; it required something more like divination. Saddam might have backup facilities already designated, but he might not. It was partially trying to predict how generals might respond to a situation they hadn't yet faced. Where *might* the Iraqi military set up a temporary headquarters if they had to improvise? What facilities might occur to the Iraqi intelligence service in the heat of battle?

Black Hole staffers began tracking down contractors who had been hired by the Iraqi government, trying to find out which structures had unique features—which ones had unique telecommunication capabilities, unique materials, or some other aspect of a building that might render it useful beyond what civilians would need.

Among the most noteworthy features of Baghdad's architecture Black Hole staff learned about was a network of bunkers. As they tried to track down blueprints and work orders, they learned something that seemed potentially significant: some of the bunkers had just been updated with reinforced concrete. Ten bunkers in particular had even been remodeled with doors designed to block electromagnetic pulses from nuclear weapons and with reinforced ceilings designed to withstand bomb strikes. The more Black Hole staff learned about them, the more they began to look like obvious candidates for backup command centers. The ten bunkers in downtown Baghdad made their way onto the wall.

4.

Chicago, Illinois

Kathy Kelly scrambled to get ready for her trip to Iraq. *What vaccines do I need? What visas?* The beautiful Kuwaiti girl was everywhere, like a challenge to her plans. On TV, in the Senate chamber, all over C-SPAN. Always in the background, hearings leaking from a TV in the other room as she packed, a picture of innocence becoming official record. Nayirah's story picked up by President Bush and repeated in speeches and on the nightly news, 40 million or so television viewers learning what Saddam was like all at the same time, as if it had been somehow coordinated.

Among those viewers, Kathy should have known, was her own mother. When Kathy trudged through a frigid winter night to say goodbye, she should have realized she was crossing into hostile territory. The new hardest thing she'd ever done, telling her mother she was going to war to try and stop it. She knew she'd be causing confusion, and maybe pain. It would seem like a suicide mission, and the risk she was taking was not hers alone. There was a cruelty to subjecting her family to this. *Not a choice, have to think of it like soldiers do, think like I don't have a choice.*

But when she came out with it, it wasn't just worry she caused in her mother—a woman who'd taken her own chances, who knew a thing or two about violence, having seen World War II close up, a nursing student arriving from Ireland to work in a London hospital while another madman terrorized a people. She wasn't squeamish about risk; she was astonished by her daughter's politics on the matter. Why on earth shouldn't we try to stop this evil man and protect his victims?

Kathy step touched at a debate—"Look at the dictators *we* support!"—but got nowhere, dropped the argument, and said goodbye. They exchanged a wooden hug, and Kathy left into a bracing wind, trailing her mother's parting words, flung out in a brogue undiluted since she'd left Ireland half a century ago and found Kathy's father in the Blitz. "But, Kathy," she pleaded. "What about the incubators? The *incubators*?!"

Kathy tried to push it from her mind. She had the blessing of complex logistics to help her block out the doubt: a train to catch, a flight from Boston, a blizzard to beat before flights were grounded. When second thoughts penetrated anyway, she tried to occupy her mind with prayer. When that failed, too, she tried poetry. Soon this would be real. She would be there, and if her country tried to go through with an invasion, she would stand there blocking the way before the tanks and troops, protecting the oppressed and vulnerable behind her.

Baghdad, Iraq

In Iraq, there was excitement. A new vibrancy to life in the cities.

Saddam's army had pillaged a wealthy neighbor in Kuwait, and it wasn't bad for everyone in Iraq. For a time, Baghdad reveled in a tide of color brought back from Kuwait: a surge of German cars, foreign watches, Italian designer clothing ornamenting the corner stores. Only when talk of American punishment grew so loud it was impossible to ignore did the mood begin to change. Saddam appeared on television urging courage. Any invasion would be weak and over quickly, he promised, but just in case, PSAs began to fill the airwaves with lessons on how to prepare for an attack that might be coming: how and where to evacuate; how to fashion gas masks from kitchen towels. The swell of novelty gave way to a rising sense of alarm leaking through neighborhoods.

Though not in all neighborhoods equally, and not for everyone. The buzz among children didn't lessen everywhere, children tending to sense the electricity passing between their parents—and some children, with their incomplete read of grown-up emotion, saw drama but absorbed only excitement. Especially children in one middle-class neighborhood called Amiriyah, home mostly to civil servants, politicians, and military officers, and to one particularly impressive building. Behind a sign that read DEPARTMENT OF CIVILIAN DEFENSE—PUBLIC SHELTER NO. 25 rose a structure the size of an entire city block. Built by foreign contractors and then renovated to contend with the better bombs in other countries' arsenals, Public Shelter No. 25 had reinforced ceilings to protect against direct hits from guided weapons, special doors that closed automatically in the event of an attack, and walls so massive and so strong that they could block the electromagnetic pulse of a nearby nuclear blast.

To children, Public Shelter No. 25 was an adventure. It was massive, a marvel, it had endless corners to explore. Three-tiered bunk beds in an open hall, a ramp going forty feet underground to run up and down, and when they tired of that, VCRs, a small video library, and generators in case city power failed.

As parents began to whisper about a possible invasion, the neighborhood started shifting its children to Public Shelter No. 25. More of them every night, to watch Bruce Lee and Clint Eastwood movies, to stay up late giggling with their friends, and giggling louder when their mothers shushed them. If the invasion happened, then even if it was worse than Saddam warned, at least the kids would be distracted, and at least here, in a building as strong as any, they'd be safe.

5.

Desert Campsite
Saudi-Iraq Border

Two hundred miles southeast of Public Shelter No. 25, Kathy and her team gathered in the desert, all of them hoping to stop a war but mostly, for now, just getting to know each other.

It was a peculiar group: a woman who claimed to be a psychic for the NYPD; a logistician who'd once been a tank driver in Lebanon; a veteran of the U.S. Treasury Department. A new "Gulf Peace Team" made up of people who didn't, at first blush, have very much in common, but who were all, in their own ways, disciples of that same teaching: willing to accept the ultimate price, the price soldiers must accept.

The plan was to land in the desert and make camp at a borderland site between Iraq and Saudi Arabia, where they knew American forces were preparing for the invasion. If the United States went through with it, Kathy and the few dozen others would be there to stand in the way, blocking the tanks and battalions of soldiers or whatever else the U.S. military was trying to move into Iraq, maybe even forcing them to turn around.

Kathy didn't know, nor did any of the others, that the attack being planned included tanks and ground troops only as a secondary element—that it would in fact be almost entirely from the air.

Still, they prepared. They waited; they kept busy. Their campsite had been built for pilgrims headed to Mecca, but it was spare and there was work to do. Digging latrines, divvying up food. Trying to escape the heat in the day and then at night, pulling their sleeping bags tight against the cold.

They boiled water to drink and brush their teeth; they played Ping-Pong in the middle of the desert; they prayed for something to happen at the United Nations, some last-minute resolution to stop the war.

Their food supply thinned. Their storage shed emptied faster than expected. There was nowhere to restock. They rationed themselves to one meal a day.

Excitement faded. Boredom set in. Then, sometimes, frustration. It grew harder to get along. They were seventy-three people from eighteen different countries, which seemed impressive, except it was hard to talk to each other. It was an ordeal to agree on anything.

Some nights, as the group was trying to agree on minutiae—could the smokers smoke at night? Could the stray cats come into the camp even though they might have fleas?—Kathy would get out of the tent and walk.

One night she realized it was darker than usual. There was no moon. She remembered something one of the activists had said: to beware of moonless nights. Pilots liked to fly then, when it was harder for enemies to see the planes. Someone pulled out a shortwave radio and they took turns huddling around it. With no news by midnight, Kathy climbed into her sleeping bag, pulled up an extra blanket, and tried to sleep. Sometime later she woke to a patter of whispers passing around the tents. She could hear urgency in the voices. Dismay, finally excitement. She heard words that sounded like, "It's starting."

She walked outside again with a blanket wrapped around her and looked up at the sky. Still totally dark, still quiet.

Then the desert erupted. Not with the sound of explosions but of *barking.* Hysterical dogs all around. She hadn't even known there were dogs nearby, but now it sounded like every dog in the world was barking itself hoarse. When she looked at the sky again, she could just begin to make out the silhouette of a jet formation, black against the black sky but catching some scrim of faint light from somewhere. Plane after plane after plane—she couldn't believe how many were up there. She felt a little shiver of horror. *If all those planes are headed to Baghdad, there won't be anything left of it.*

6.

Downtown Baghdad, Iraq

Just before 3 a.m. on January 17, 1991, Weldon Word's latest creation initiated Operation Desert Storm. A new Texas Instruments Paveway III fell from a new Lockheed stealth bomber and glided toward a tower used by the Iraqi telecommunication ministry. When it was almost directly overhead, the warhead pitched its nose down, picked up momentum, and struck the tower's roof with such tremendous downward force that it crashed halfway through the building to the middle floors before detonating.

At almost exactly the same time, Paveways flew toward electrical substations. Baghdad went dark and then lit up again with barrage fire. Iraqi forces couldn't see what they were shooting at, but flak and SAMs crisscrossed the sky. Speeding, glowing tracers wove back and forth like sprinkler spray while nearly seven hundred U.S. military planes from aircraft carriers and bases surrounding Iraq scored direct hit after direct hit. The sky was alive with planes and bombs and guns trying to shoot them down, but at the American bases squadron leaders kept landing and reporting in: *Targets struck, everything on schedule.*

A satellite uplink from a hotel in Baghdad piped war correspondents into a studio in Washington, D.C., where a retired Walter Cronkite had come back to the anchor's desk to sit next to Dan Rather and help the world process the revolution unfolding on-screen, Cronkite giving a private pep talk as millions listened in. "Don't grandstand this one," Cronkite told a young reporter calling in from Iraq. "Why, you know, save your skin, boy." The young reporter crouching in the Baghdad hotel paused, as if blindsided by this bit of sentimentality from a deity of the profession. "This is just the greatest story in the world at this time," he finally managed, "and I'd like to cover it as much as I can." Outside, tracer rounds colored the sky, staccato pulses flashed the ground to green, and a rising tempo of explosions shot journalists' voices into a more urgent key.

Back home, bundled inside against the Chicago winter, Kathy's mother watched the war play out on TV along with much of America and some of the world. Angry squinting orange rings on the backs of fighter jets vaulting off aircraft carriers, black-and-white crosshairs over nighttime buildings as they flashed to smoke. She learned the war was going well. Laser-guided bombs were destroying Saddam's telecommunication centers and paralyzing his capital.

She watched an evening news gone necrotic, green hours of flashes from featureless buildings, night-vision vengeance for all those babies thrown onto hospital floors. When else had the mighty been so righteous? The world had righted itself since Hitler's indiscriminate terror drove whole cities underground; now it was the murderous dictators receiving the punishment of incalculable power, only this time that power was discerning. Only the deserving few were struck. In America, a country was comforted by its own superior strength and the sense that a world would be put at ease with this proof that the most powerful military in the history of the world was also its most just. Deployed in service of the weak and setting an example: that no dictator could hurt babies without reaping America's precise and righteous might.

Desert Campsite
Saudi-Iraq Border

Out in the desert, closer to all of it than anyone watching the broadcasts, Kathy Kelly had almost no idea what was happening. She had no access to TV, and what team members could pick up on the shortwave radio gave little sense of what was going on. Whatever Baghdad was experiencing, out here at their pilgrim campsite, nothing happened. The Peace Team played Ping-Pong. Rations dwindled further. They'd stopped no violence. Maybe they'd find a way soon.

Warplanes passed overhead. With little else to do, Kathy started counting them: at night, a new formation every five minutes.

After sixteen days in the desert, a tired-looking Iraqi man arrived, leading a small convoy of school buses, saying he came from the Ministry of Culture. He asked who the leader was. They told him there wasn't one: "We've arranged ourselves into affinity groups." He looked puzzled. He addressed all of them: he said there was now word that the American military was planning a pincer movement to follow all the bombing. There would finally be a ground operation, and it might come right through this campsite. He'd been ordered to bring them to a hotel in Baghdad, where they'd be safer.

Members of the Peace Team began to argue with one another. Most didn't want to leave. Some wondered how Baghdad could possibly be safer when they saw so many bombers flying in that direction every night. Others wondered why they'd even consider leaving: if a ground invasion really was coming through this site, that just meant they finally had the chance to do what they'd come to do.

But with supplies nearly tapped out, they wouldn't survive in the desert much longer anyway. And this was a man speaking on behalf of the people the Peace Team was there to protect, now nearly begging them to leave. So, finally, they acquiesced. Maybe once they got to Baghdad, Kathy figured, they'd find a way to be of use.

As the buses pulled out and Kathy encountered the first real signs of war on the highway into Baghdad, it started to feel strange that they hadn't gone there in the first place. They'd figured most of the violence would be over land, and so far almost none of it had been.

But now the signs of destruction were everywhere. The school bus swerved around cracked blacktop, potholes from recent bombs. They passed blackened oil tankers still leaking inky coils of smoke. She saw burned-out ambulances. Shells of civilian cars burned the color of rust, and the skeletons of other buses surely full of passengers who had tried to escape.

Kathy pressed her face to the window. It was quiet inside. People had died out there, here, on this road. There were maybe still bodies in the flipped-over cars she was passing.

But she felt OK. She found she could keep *feeling* pressed back, at arm's length. She was a visitor passing through a museum with relics that were real and disturbing but safely ensconced behind glass. She found she could handle it.

When the buses downshifted and pulled off the road to a rest stop for a bathroom break, the team found a different kind of damage: a rest stop that hadn't been staffed since the first days of the war. Rivers of shit overflowed; nothing had been cleaned. These were signs of a society no longer able to keep the things humans do belowground and out of sight.

Still, Kathy felt she could manage. She'd survived first contact with war; she was tested now. She felt ready for whatever she had to do next.

The buses pulled back onto the highway for the rest of the trip to Baghdad.

IN BAGHDAD, INSIDE Public Shelter No. 25, as the war raging outside reached its third week, a group of mothers decided the children could use a distraction. They were spending more time in the shelter; they'd been disconnected from their regular lives and had seen little of their fathers, since even as big as the shelter was, there wasn't room for everyone, and the men were largely forbidden. Sometimes the men would drive by to visit. But it was a time of separated families, the city was in the midst of near constant bombardment, and the mothers agreed a gesture at normalcy might be a salve for some of the fear that now gripped them all.

The Eid holiday was coming up, an excuse to celebrate. And anyway, they were living in one of the strongest buildings in the entire country. They'd been told they were safe here inside Public Shelter No. 25. Regardless of what happened out there, the walls around them were strong enough to protect against even a nuclear attack.

It didn't figure, as they began gathering ingredients, that they should worry about what might happen when a building designed to withstand a massive explosion outside was subjected to one from within.

Or that, as they were planning their celebration, Saddam's regime was making a key strategic decision that could actually put them in greater danger. The regime was clearly losing control. Its command-and-control ability was almost totally degraded. They now had reason to fear losing the support of their own people, who were increasingly disconnected from the government and now, having witnessed America's awesome airpower for weeks, had less reason for reverence to Saddam. So, running out of ways to prevent entirely losing control of its people, the regime had turned its attention to one of the few factors preventing complete American psychological authority over the Iraqi population: outrage at American-caused civilian casualties.

7.

Black Hole Planning Cell
CENTCOM

In Saudi Arabia, in their annex under CENTCOM, Black Hole staffers marked progress on the wall. Reports came in, one after another: Saddam's telecommunication network was offline; another part of the power grid had collapsed; another node of his security services had suffered a direct strike. So many bombs striking targets on a widening circle of Saddam's power structure that soon, per Colonel Warden's charge to prevent the unnecessary casualties from letting the enemy retrench and extend the war, they began to refocus on buildings that could serve as backup facilities. When American bombs destroyed a complex near the Baghdad Equestrian Club that served as the headquarters for Saddam's secret police, Black Hole planners were already able to predict where it might relocate.

One structure in particular matched the parameters a relocating intelligence service would require: a block-sized building in a middle-class neighborhood called Amiriyah, a building that was physically imposing, easily big enough to house an intelligence service. Overhead photos showed the roof was painted in a camouflage pattern, and they'd learned from CIA sources that the building had recently been retrofitted with reinforced concrete and massive doors that would close automatically in the event of a nuclear strike nearby.

Once Black Hole staffers began looking at Public Shelter No. 25, the evidence seemed more and more damning. They noticed military and other government vehicles coming and going and though there was a sign in front that read DEPARTMENT OF CIVILIAN DEFENSE—PUBLIC SHELTER NO. 25 in both English and Arabic, that could be easily dismissed. Of course signs would claim civilian use, even if—*especially* if—it was being used for military purposes. This was not a *shelter*. It was a bunker for Saddam's security services, and it needed to be eliminated.

One general looking over the evidence dissented. Maybe the roof *was* camouflaged. *Everything* in Baghdad had camouflage. Even the water treatment plants had camouflage.

And maybe military vehicles *were* coming and going. It was war; military vehicles were coming and going everywhere.

But just as the hesitant general was convincing Black Hole staff to take Public Shelter No. 25 off the target list, several highly placed officials in the United States government received sealed envelopes containing a fresh intelligence report from an asset inside Saddam's regime: a double agent, reportedly close to Saddam. The report said that Saddam's General Intelligence Department had relocated to Public Shelter No. 25, just as the American spy agencies suspected.

The DIA, CIA, Black Hole, and CHECKMATE at the Pentagon reached a consensus: Public Shelter No. 25 was a backup command post, and time was running out to strike. The master attack plan was updated. Munition and aircraft were chosen: two stealth bombers, each armed with brand-new "bunker buster" Paveway IIIs. A slot for the mission in the overhead orchestra of planes coming and going was found in the 1:30 to 1:55 Zulu block, 4:30 to 4:55 in the morning local time, on February 13.

Grid coordinates for the target were written into the attack plan, as was a brief target description:

> Activated, recently camouflaged command-and-control bunker.

8.

King Khalid Air Base
Operating Base of United States Air Force 37th Fighter Wing
Southwest Saudi Arabia

At an air base wedged into the mountains in Saudi Arabia, ground crews wheeled the new Paveway IIIs out to two stealth F-117A Nighthawks, call signs Heat 34 and Rain 37, and levered the weapons up and into the open bomb bays.

They secured the bombs to their mounts and pulled the arming pins out. The warheads were set to delayed fuse mode so they wouldn't detonate until deep inside the target bunker and, if all went as planned, close enough to any oxygen or fuel tanks the bunker had to trigger secondary explosions from within.

The bombers' downward-facing doors swung closed, the bombs now invisible behind the radar-absorbing skin.

Two pilots with mission information clipped to kneeboards climbed ladders into the cockpits. Each pilot dialed the target's longitude and latitude into the cockpit navigation system. They each radioed in turn, asking ground crews to pull the cords still tethering the aircraft to the ground.

At around midnight, both planes taxied, took off, and turned northward in a trail formation, one pilot following the other, flying mostly in silence.

When they had to correct their heading, they did so with wings level, turning without banking to avoid showing a more radar-visible surface.

The lead pilot signaled for an ops check.

Both watched their displays to check the jet engines, instruments, altitude, and oxygen systems. They compared the fuel left in their tanks with the kneeboard charts showing the fuel they were supposed to have remaining.

The two stealth bombers rendezvoused with an air tanker to refuel.

They detached, continuing in silence until, from several miles out, the city crept onto the screens embedded in their instrument panels, appearing as splashes of heat from the infrared camera.

Through the cockpit windows there was still little to see, but on-screen the leaky splashes of light and dark resolved into avenues, streams, the river. Clumps separated and became compounds, compounds became individual structures, then streets, homes, backyards.

Heat 34 and Rain 37 arrived over southwest Baghdad at 4:30 a.m. The lead pilot keyed a display to show armaments. The release system was now prepared. He switched to WEAPONS ARM—OFF SAFE.

He checked the coordinates on display. Correct altitude, correct airspeed, correct heading. He turned the MASTER ARM switch to ARM.

Now the screen showed images from a different infrared camera, this one pointed directly down. The ground rolling below, a block-sized bunker moving toward its center.

The pilot checked the photograph strapped to his kneeboard. It seemed to match the building on-screen. "Target acquired," he said, then clicked off, still careful to keep comms to a minimum. A subtle move of his fingertip against a button on his joystick moved crosshairs on the monitor. He centered the crosshairs directly over the aimpoint, a ventilator shaft on the roof of the bunker.

He pushed a button to turn the laser on. Slaved to the crosshairs, the laser now shone through miles of sky toward a spot on the roof of the shelter, down the ventilator shaft, and onto a spot deep inside the building, where five hundred people slept off their holiday meal, oblivious to the laser reflection, invisible to the human eye, sparkling nearby.

The pilot no longer had to look down, a computer now projecting symbols up onto the windscreen, telling him he was within Paveway's range. He pressed the CONSENT button and the weapon bay doors opened.

He heard a clunk, the plane getting a little lighter.

The weapon bay doors closed.

ONCE SEVERAL FEET below the plane, the Paveway III's fins extended. Its thermal battery woke up and began firing electrons across circuit boards. The weapon's brain processed a course through the sky, searching for the laser reflection but also keeping its body aligned just so, making subtle adjustments, aiming the hardened warhead not directly to the target but to a spot just over it, so it could nose down and drop almost vertically, gathering up all its momentum to defeat the bunker's reinforced roof and destroy whatever lay inside.

The pilot in the other plane pressed his CONSENT button.

Inside the shelter, an enormous crash woke the children. A 2,000-pound warhead smashed through the ventilator shaft, crashed through the top levels of reinforced concrete and multiple floors housing women and children, a second followed down the path carved by the first, and a fraction of a second later, the bombs detonated.

The pilots made a programmed course alteration and headed home, while the shelter doors swung shut automatically, sealing the people inside as the temperature rose to nine hundred degrees.

9.

The Al Rashid Hotel
Downtown Baghdad

A breeze eased against Kathy's hotel window five miles away. A rattling of glass, in minor argument with something out there. Another bomb somewhere. Every time that window moved, a sea sloshed around in her chest. *We failed.*

She took a pillow out to sleep in the hall, couldn't sleep, and got up to find someplace to write in her journal. By now there was only occasional power. She groped her way down the hall, dark save for sporadic flashes from the cracks under hotel room doors. She moved slowly, running her fingers along the wall until she found a surface that gave, pushed harder, and felt a door open. She could tell by the echo of her breath that she was now standing in a bathroom. Her fingers found a light switch and she tried it just in case. The bulb must have had its own power source, because the light flicked on and she was no longer alone: a clutch of huddled women and children blinked up at her.

One of the women twitched just enough for Kathy to understand a message being transmitted: *Please don't show fear.* If Kathy looked frightened, the children would know to be frightened. She tilted her head, pushed worry off her face, and, standing there playacting in the sudden light, understood something new about herself: she could put her mind someplace else, someplace safe, far from the violence. That must have been why she could see all the destruction on the road and keep steady. She could keep from breaking down, she could disassociate, and this could be her own weapon against the weapons. She had not been able to stop a war yet, she had not yet prevented any bombs from dropping, but now she knew she had this power to fight this artifact, the infection of trauma. Even just by looking at the children and keeping steady.

When a bomb landed in the parking lot across from the hotel, the man from the Ministry of Culture came to tell Kathy he now had orders to get the foreigners back out of the city, this time across the border to Jordan.

And once there, Kathy learned it was Jordan that aid trucks were leaving from, but that it was a dangerous trip since there were bombers overhead. She and the team devised a way to help. Kathy called the American ambassador in Jordan to tell him they would be in the trucks going back and forth to Iraq. Perhaps team members could provide comfort to the drivers, help with loading and unloading, maybe just company.

"I'm going to have to tell you, do not do this," the ambassador said. "It's really unsafe."

"It's unsafe for everybody on that road."

He sighed and changed his tone. "Well then, I have to put on my other hat. I'm going to have to tell you if you carry on with this, you risk being prosecuted back in the United States."

The threat didn't bother her. For the next week she and the Peace Team members helped the truck drivers assemble supplies and joined them on their runs back and forth to Iraq. When someone pointed out how lucky it was that, even on a road known to be a prime target for American bombers, none of the trucks had been attacked, the team realized it had proven a new tactic. This war was being fought almost entirely from the air. Maybe that was the way war was going. If militaries were shifting tactics, the Peace Team would have to as well. It was no longer a matter of trying to block armies from moving but of stopping bombs from falling. And maybe the way to stop bombs falling from the air was to be there, on the ground, riding alongside the target.

Embedding with the convoys allowed team members to make other observations about how war looked from the ground. She found herself traveling with Red Crescent volunteers, doctors and nurses, and back in Baghdad she found a spare cot on the side of a hospital room. She watched hospital staff try to keep people alive without city power and dwindling fuel for generators. Medicine spoiled because of limited refrigeration. Vaccines had to be discarded. Babies were dying of exposure when sealant for old incubators failed. Doctors tried to make do without them, telling new mothers to hold premature babies between their breasts to keep them warm and showing them how to jury-rig incubators with cellophane wrap. Kathy thought of Nayirah in the halls of Congress, asking for America's help defending Kuwait against Saddam. Now, in front of Kathy, babies were dying, but here they weren't dying because of Saddam. They were dying, she thought, because of *us*.

With scant access to news, Kathy only intermittently became aware, from newly arriving Peace Team members, that back home reporters were beginning to unravel Nayirah's story. The pretty Kuwaiti girl hadn't just shown up by chance; she'd been part of a sophisticated lobbying effort by a company advising that, for Americans, tales of atrocities against civilians were sure to gin up support for the war. It turned out that Nayirah was not a chance bystander: she was the Kuwaiti ambassador's daughter, coached and scripted by a public relations company that had not just prepared her testimony and helped her practice it but took the footage and distributed it to seven hundred local and regional television stations. To secure American support for the war, Kathy eventually learned, Kuwait had paid millions of dollars.

And it was also only from one of the newly arriving Peace Team members that Kathy learned, a little too late, about a middle-class neighborhood called Amiriyah, just a few miles from the hospital the team had turned into its latest outpost, a neighborhood with a building where a new kind of American bomb had defeated reinforced concrete and exploded alongside sheltering civilians.

Kathy had a sense that members of the team needed to go there. She poked her head out into the corridor, stopped a scrum of overworked Red Crescent workers, and asked if one of them might arrange a vehicle.

Amiriyah Neighborhood, Southwest Baghdad

The first thing that Kathy noticed about the Amiriyah neighborhood was how orderly it was.

Looking through the truck window as the driver pulled in, the houses seemed almost disciplined. Everywhere else in the city, there were weeks' worth of gathered trash, but here, in Amiriyah, the streets seemed exceptionally clean. When she got out, the neighborhood struck her as *dignified*, each home in its place. Brick houses set behind walls, but low walls. Walls the right height for decorum, not high enough to comment on solitude, and something in her flickered and awoke. These were people who shared recipes with one another. Who borrowed spices and looked after each other's children. She could have been looking at a lightly faded photo of her own home.

Except that here, above her, big black banners hung from upper windows with streams of rolling white script, Arabic lettering she couldn't read but still knew were the names of the dead.

She followed the banners, one after the other, down the rows of orderly houses until she was looking at a massive square structure where fire had burned for a day and desperate parents trying to reach the children inside kept getting pushed back by the heat. Another thing she now understood: parents had *delivered* their children there, and many had then left in order not to take up room. While they were gone, bombs crashed through a ventilation shaft, an industrial water heater exploded, and children trying to get out ran into blast doors slamming shut and sealing them in, the whole structure becoming an oven with boiling water racing past their legs. Parents had delivered their children here and then didn't see them again until the inferno had tired itself out enough for firemen with acetylene torches to cut their way in and pull bodies from the rubble by then unrecognizable even as bodies, burned down to fragments that seemed too small even to be children.

Kathy found herself standing inside. Inside, even without electricity, the building was illuminated. She looked up. At the bottom of the ventilation shaft where the reinforced concrete had given way to a perfectly angled bomb, steel rebar splashed outward like a forceps holding open a wound, and an assault of sunlight shone through, filling the space with enough light to see ribbons and photos of the dead already beginning to accumulate. Enough light to make the observation, strange even to herself as she made it, that the flowers were probably plastic.

As she walked back out into the middle of the neighborhood, the fact that assaulted her was that children had been scalded to death in the one place they were supposed to be safe. By the number of black banners hanging from windows, there must have been hundreds of children inside when the bombs fell, and now, standing in this altered suburb, after weeks of keeping her composure, of learning she could disassociate when she needed to, she found that here she no longer could. She began to cry, for the first time since arriving in the country, and as all the accumulated weight of the previous weeks finally landed on her, she felt a strange pressure around her waist.

She looked down and saw a tiny girl standing there. A striking child looking up, arms wrapped around Kathy's body. The child smiled.

In English, in a squeaky little voice, the girl said, "Welcome."

IN TEXAS, WELDON Word was moving through a new phase of life, feeling something new for him too.

He'd decided he'd reached as good a moment as any to step back from TI. Disciples of his were now filling influential positions throughout the company. People who'd worked on Paveway and absorbed Weldon's vision were now executives and board members. Paveway's legacy felt secure, and he'd begun to think about his own future apart from it. His wife had gotten involved in rural real estate, finding second homes for clients looking for weekend escapes from Dallas, and the two of them decided they might as well make the migration as well. They shifted into a quieter life out in Tyler, a place known less for tech and weapons; a place whose climate and soil were ideally suited for, of all things, roses. After a day of DIY home improvement in the "Rose Capital of America," Weldon came inside, turned on the TV, and saw the Desert Storm footage much of America was seeing.

But while other Americans saw footage of almost a clean war, where visible injuries were mostly to buildings, Weldon saw something different.

He'd always known he was building weapons, but he'd never had to think much about killing. The *saving* had always been more front of mind. Ever since Joe Kennedy's ill-fated mission amid the ensemble of men across the century grasping for a "steerable bomb," the impetus was usually the pursuit of safety, not destruction. Weldon had started on a project to give ground crews a tool that was easy and safe to work with, and pilots a tool that was a little safer for them to use. He'd had the best, most careful people work on its assembly lines. The only time he'd actually seen the bombs deployed was on the test range—where they were often inert, filled with instruments instead of explosives—and in the little footage they had from that one-night raid in Libya during El Dorado Canyon.

Now that one-night raid in Libya played out every night, and Weldon could tell, even from the sanitized Pentagon-issued targeting pod videos, when the bomb being shown was Paveway. Off on its own, steering toward the important parts of a city and threatening the people who brought it into the world with the sense that they might not just be inventors, not just creators, but destroyers.

Weldon tried to distract himself with the notion that this war was justified, and if you had to fight a war, the best way to save lives was to get it over with quickly. Paveway could still be something that, in the end, spared lives.

But it didn't entirely settle in. He went on a walk. He climbed a hilltop near his home by himself, and he prayed.

BOOK V

RADICA

(OPERATION ALLIED FORCE)

1.

As the separation between Weldon Word and his most influential creation widened, another early idea of his was beginning to gain traction.

A few years before the Soviet Union collapsed, he'd sat down with TI executives and laid out one of his now-trademark extravagant visions: he said he thought there could be a way for a bomb to navigate not just by laser, but by using information coming from satellites in space.

Working on the bunker buster Paveway III for the stealth bomber had given him a window to another version of the future. He saw the shift in tactics a successful stealth program might inspire. A bomber invisible to radar could shift preferences back to high-altitude bombing. High-altitude bombing brought one of Paveway's key weaknesses back into play: it was too easily foiled by weather. Ever since the days of Rick Hilton orbiting over a bridge, waiting for a break in the clouds so his back seater could see the target, Paveway had always been limited by the fact that it couldn't see through obstructions any better than the human eye. If anything, it probably saw worse: water molecules in clouds could scatter the laser beam into a useless spray of refracted light, and dust or smoke could block it altogether, rendering the bomb effectively blind and unguided.

For Paveway to achieve its full high-altitude potential, Weldon needed to fix that. Paveway needed some technology less susceptible to clouds and the debris flung up by other bombs and anything else going on below the plane.

Perhaps, Weldon figured, the answer was something *above* the plane.

He'd started thinking about satellites. A kind of satellite capable not just of taking photographs but of transmitting other kinds of information that might help a bomb find its way.

To TI executives, the idea had seemed even further away than Weldon's earlier bets on over-the-horizon markets, potentially even more expensive, and even harder than usual to envision as a viable product. Early prototypes of satellite "positioning" systems cost hundreds of thousands of dollars; were so large that they barely fit on a cargo ship, let alone inside a bomb; and their geographical coverage was too limited to be all that useful. To use satellites for something like navigation, there would have to be enough of them in space to service any part of the world users might find themselves in. To build a system useful for navigation, it would have to be global. There were far too few data satellites in orbit for that.

To TI execs, it wasn't just that there was no guarantee Weldon would actually figure out how to make navigation by satellite work, or that it would make the company any money. Even if Weldon's team devised a usable way of helping someone or something find its position in space, for a truly "global" positioning system—or, for short, "GPS"—TI would be dependent on too many forces outside of its own control. They'd be dependent on an Air Force program to build enough satellites for the system to work, and on NASA, which was responsible for launching satellites.

But, far-fetched as it was, Weldon's vision happened to align him with that of another TI living legend: Pat Haggerty, the World War II Navy vet who had first convinced the company to buy into something no one really understood called a transistor. That investment had led to the silicon semiconductor chip, the integrated circuit, and then TI's edge both in smart bombs and personal computing.

This time, it wasn't experience in the Navy that Haggerty was drawing on, but a television show. Haggerty had been watching a series called *Cannon*, about a private investigator who drove around in a car that had a telephone unconnected by cable to any network. The phone in *Cannon* was wireless, so the character could talk from anywhere.

Pat Haggerty figured the technology required to build a communication device like the one imagined in *Cannon* was probably not as far away—or didn't *need* to be as far away—as most viewers thought. And if someday people were driving around with communication devices that were wireless and mobile, there was probably a way GPS could add value to that package.

With Haggerty on his side, Weldon had been able to argue that TI could do for GPS what it had done for personal computing: it could use weapons as a package to sell a new technology to a rich enough customer, in a high enough volume, that they'd bring the price down and seed a civilian market too. Plus, investing in "GPS" now would help TI protect its turf as the leader in smart weapons, because at some point, however far in the future, *someone* was bound to find a way to guide a bomb by satellite.

Together, the arguments worked, and a familiar pattern played out. A meeting that began with a room full of skeptical TI executives skittish about funding another Weldon Word flight of fancy shifted so completely that at the end the head of strategy turned to the company's CEO and said, "Weldon didn't ask you for enough money. You gotta give him more."

Once again Weldon's team poured itself into the task of realizing an over-the-horizon technology, and once again, it hadn't taken long for serious obstacles to emerge.

Just as TI feared, the Air Force struggled to get satellites up and running. Funding cuts slowed satellite production, and workmanship problems plagued a line of flight boxes coming off a plant in California, further delaying the advent of a truly global satellite network.

Then, in January of 1986, disaster struck. The nation watched in horror as the space shuttle *Challenger* exploded on live television, killing all seven astronauts aboard, including a schoolteacher. As a traumatized nation mourned, obscured amid the tragedy was the fact that, along with the astronauts, *Challenger* had been carrying GPS satellites for the Air Force, and the Air Force had been depending on future *Challenger* flights to finish delivering the network of eighteen satellites necessary for global coverage.

After *Challenger*, the notion of GPS was pushed further into the future. TI kept working on receivers, but with the satellite network delayed indefinitely, so, too, was the full promise of weapons—or drivers, or pedestrians—using GPS to reach their destinations. By Desert Storm, the military was only just beginning to find ways to use a limited version of GPS in combat, and the effort to add the technology to Paveway would soon meet with even more trouble, as the team was about to face perhaps its biggest crisis to date.

2.

Somewhere in the North Adriatic Sea

It's December, just before Christmas, and Kathy Kelly has found herself on the upper deck of ship, sailing through a storm.

On the lower deck, a few hundred seasick Italians stumble over themselves as waves break over the railing and vomit spreads on the decks. It's chaos down there. It feels in this moment like chaos everywhere, the early days of a new world order.

Just months after she returned from Iraq, the world's gravity changed. The Soviet Union collapsed, unleashing a surge of elation in the West, a brief reprieve of clearing skies. But the Cold War had been an organizing principle for its participants, for the entire world really, and it was an arithmetic that suddenly no longer applied. As communists lost dominance all over the world, shoots of long-suppressed identity broke through, and countries assembled out of different kinds of people began to pull apart.

She's on her way to Sarajevo, a besieged city on a landmass trying to reshape itself amid that shifting logic. Yugoslavia, a country that by now, in 1992, is becoming a Colosseum of teams fighting against yesterday's friends and neighbors. Mothers shared meals one day and the next watched their sons stop playing soccer together and disappear into the mountains to join militias canted against one another. It wasn't just families splitting apart but neighborhoods, villages, entire cities. As if the boys going up into the hills pulled cords to drag their families partway with them, and in the growing space between neighbors more permanent hatred could sluice in and harden.

A group of mostly Italians watching the conflict from afar had wanted to do something about it. They learned about the Peace Team in Iraq and called Kathy to ask for her help doing a similar thing again.

This time they have name tags; they've brought vehicles. Three fully stocked ambulances and a truck full of humanitarian supplies wait in the hold of the ship.

This time there are more people too, nearly five hundred, mostly Italians and a few joining from Belgium, from Germany and the United States, all looking more unified, more organized. But the night before they left there was no planning left to do, only worrying, and though the Italians had passed Father Berrigan's test—they were willing to pay the Ultimate Price—they were perhaps not ready to do so sober. They'd been up all night dancing to a bagpiper and drinking themselves a little distance from what they were about to do, then woke up aboard a ship headed through a once-in-a-decade storm.

Kathy tried to bring sick bags to the worst off, but there were too many of them and too much to clean, and eventually she just gave up, leaving them to empty their stomachs while she climbed to the upper deck, where she now tries to enjoy the storm.

THEIR PLAN IS half a plan: to dock on the western edge of the splitting-apart territory and ride inland another hundred miles to Sarajevo, the once-vibrant city now being starved, strangled, and mortared. Of all Yugoslavia's breakaway republics, the conflict seems to be reaching a peak in Bosnia, and the siege of its capital, Sarajevo, was beginning to look less like one party trying to defeat another and more like ethnic cleansing.

The Yugoslav military, mostly made up of Serbs, has encircled Sarajevo, an almost entirely Muslim city already pinned between mountains. The military blocks food from being taken into Sarajevo, has cut power and water, and places snipers on walls at the city's outskirts to fire at people trying to make it to the hospital on foot.

Kathy feels the new tactic—putting themselves under the weapons—may not work here. Here, the weapons being launched aren't being launched by her country. Here, stopping the conflict may require *attracting* American attention.

THE TEAM ENDURES the storm without serious injury and docks on the western end of the crumbling country. The next morning, Kathy and the Italians board buses for the overland part of the trip to Sarajevo, snaking inland on narrow roads and over snowcapped mountains. They ride through small towns of small buildings with shot-through windows, haphazard weapon spray beginning to focus as they crawl east. Through towns with names Kathy can't pronounce: Kiseljak, Posušje, Lištica. A place called Mostar, which someone says means "the Old Bridge," for the old bridge that connects two halves of it.

The buses climb switchback roads so steep they're stacked almost vertically, and Kathy feels the transmission grabbing at its lowest gears, lurching forward, until finally they rise above the snowline and the path flattens out just as night falls.

They find an abandoned school, break in, and arrange themselves on the floor.

Only twenty miles from the siege now. Kathy closes her eyes.

A second later she's wide awake: an ambush, people all around her, a mass of blank eyes blinking down. It takes a foggy moment to realize that none of the intruders have weapons, that all of them are oddly small in stature—that they're actually all children. Another beat to register that the team didn't find the school abandoned because of war; they found it abandoned because they arrived at night. Now it's morning and the students have shown up for class to find five hundred Italians splayed out and snoring and one small, curly-haired woman rising to negotiate an armistice.

The youngest of the students adapt quickly, disappearing and returning an hour later with cigarettes and snacks to sell, making change as the activists roll up their sleeping bags and file into the school's gymnasium for a last brief on how to survive the violence they might soon encounter. *Keep your mouth open so exploding shells don't puncture your eardrums*; *if a sniper hits one of you, don't go to help, take cover, because snipers sometimes shoot whoever comes to help; if a mortar falls, run to the crater it made, because the odds of two mortar shells landing in the exact same place are low.*

Then back onto the buses to pick through difficult terrain for another day, finding another school to make camp in—one that's actually abandoned. Kathy sleeps in spurts, because the call and response of machine-gun fire keeps bouncing off the mountains.

One more day inching east, another school to camp in, another strange visit. This time a grown man, all alone, ambling in to greet these visitors and asking if he might take Kathy on a tour of Sarajevo. An odd offer, but Kathy likes the idea of seeing the siege with someone who knows the city. And for the moment at least the machine guns and the mortars have quieted, militia members making their morning commute from their firing positions to their day jobs, and anyway, she soon finds herself so engrossed in conversation with the man that she forgets to worry about snipers at all.

She finds this man charming. She finds his pride in the city charming. How many cultures it holds, the mixing of peoples, even as they walk through streets cluttered with concrete from rockets one people had fired at another.

It's the two of them in their own cold dreamscape. The city feels entirely empty of anyone else. When the odd car passes, it's redlining to avoid sniper fire, there and gone in a loud tear of a moment, sealed away so it feels imagined. Then just the two of them again.

They pass homemade shields, skinny graffitied arrows scarred onto walls and pointing toward sniper nests, the only trace elements of human life, until they come across a mirage: a group of young people huddled around a café table, drinking, as if oblivious to the crumbling buildings around them. Like actors on a movie set taking their break, assured that the violence is all pretend.

Otherwise, the city is all hunger and destruction, and she still can't see a way to cure it. But no mortars have fallen during her walk and the hills have been quiet, so maybe whatever she's doing is working. Her wild hair a flare to militias in the mountains, advertising the presence of a foreigner whose death might cause them more problems than it's worth; more problems than a Muslim's.

She decides to invite her new friend to join a march through the city that the activists have planned, an interfaith procession, the five hundred of them hoping to draw people out to join their peace songs.

It's intended to be a reminder that people can come together, but soon after they start marching, they have to stop suddenly: a woman shoots out from somewhere among the rubble, runs into the middle of the street, stops, and stares them down, blocking their path.

Other women follow her, and soon a whole huddle of them stands in the activists' path. Staring down the marchers; just watching.

And suddenly Kathy understands: the mass of activists, hundreds of them young men with Mediterranean features—they must look like the sons who've left these mothers to fight. A procession of returning ghosts, and the women, after a moment of standoff, begin to throw their arms around the marchers. Holding them by the necks, weeping, pretending. And together, with new reinforcements, the march begins to inch forward again.

They sing their songs at indifferent buildings, hold their candles and sway with the music, and they try to remember the words to a song people sang during the Olympics that were held here, somehow, only eight years ago.

And that's it. They've made their point. They've walked around; they've borne witness. They hope their visit raises awareness back home of what's happening here. They've been here for a total of two days—two days during which no bombs have fallen—but there's nothing else they can think of to do besides head home, so they head home. Leaving behind the truck full of humanitarian supplies, boarding the buses, passing back over the snowcapped ridges on near-vertical switchbacks, back through the places with mostly unpronounceable names: Lištica, Posušje, Kiseljak, Mostar, "the Old Bridge."

And it's on the route through Mostar that Kathy and the team drive past a massive, state-of-the-art weapons facility hidden just out of sight, just on the other side of a wooded area, and just then in the midst of a transition.

3.

Military Technical Institute
Six miles north of the city of Mostar
Bosnia and Herzegovina

The facility on the way out of Mostar housed a formidable arsenal. Everything needed to mount a chemical and biological weapons program, from kennels and breeding stations to raise animals for testing, to a manufacturing center that could produce choking agents at a rate of 20 kilograms a day. The Mostar facility produced VX nerve agent and nitrogen mustard gas and had forty metric tons of sarin gas precursor on hand along with 250 artillery shells already loaded with sarin and ready for use. Rockets compatible with a truck-mounted multiple rocket launch system were armed with chemical agents as well, so a single truck could fire a dozen nine-hundred-pound poison-filled rounds at a time.

As activists tried to draw the world's attention to the conflict here, Serb forces with control over the facility worried that it might be found. They began quietly breaking down equipment, sealing the chemical weapons into special containers, loading them all into trucks and sending them east, to a place where the arsenal might escape detection.

The moving arsenal did not escape attention. American analysts examining satellite imagery tracked the trucks moving farther and farther inland, following them until they stopped and began to unload, 120 miles east, in a smaller factory, in a small, forgettable, inland city called Lučani.

Lučani, Serbia
Former Yugoslavia

In Lučani, one young mother and her daughter went about their days with little sense that world events were bearing down on them.

The girl, Kiki, was a child for whom the notion that there were things out there, up there, that could hurt her didn't figure. She was young and unsinkable.

Her mother, Radica, wasn't much different. She understood, in a general way, that things had been bad elsewhere in the world. But here, with a daughter, a husband living in a different city, and her job hosting a radio show helping other young mothers over the air, there was little time to mind the fractures rattling down the peninsula.

She knew there was a siege in Sarajevo, but that was a four-hour drive from here. Over mountains and valleys—a drive she'd never had a reason to make.

She knew there was violence there, but the violence—between a militia that claimed to be fighting on behalf of Serbs like her, and the militias fighting against Serbs like her—was not happening here, not in her little anonymous town. Not in Lučani.

Here, there were *only* her people. The anger against Serbs elsewhere was hard to fathom. Here was an unmixed enclave spared the combustion happening valleys away.

Anyway, here she was busy with her radio show, and with the task of parenting mostly alone. She had her hands full trying to keep track of her daughter, a curious girl who had a habit of wandering.

Especially when there was something new to see.

4.

Washington, D.C.

From the White House, the siege of Sarajevo looked like something from a bygone era, from the time of knights and horses, but President Bill Clinton was young and confident enough to believe he could do something about it.

The crisis was far away, the fighting parties entrenched, but he'd come into office already bucking conventional wisdom.

After the televised success of Desert Storm, President Bush's second term was virtually assured. The Cold War was won, Bush credentialed as a wartime president, and when the new crisis overseas began, there was little upside in him wading in. The Siege of Sarajevo was part of a conflict so poorly understood that journalists routinely confused the parties involved, and the term for the process underway there—"Balkanization"—seemed to just mean chaos.

Bush's approval ratings remained so high as the campaign season approached that the Democrats had a hard time finding a viable candidate willing to challenge him. Among the few people ambitious enough to try was the young governor from Arkansas, who decided to launch a quixotic run and who, with little to lose, told the few reporters willing to listen that he thought the Balkans deserved American attention. Next to Bush, Clinton started to sound almost Churchillian. And as more coverage emerged from the Balkans—children killed by snipers, emaciated bodies at concentration camps—President Bush's restraint began to look less like maturity, more like a lack of resolve, and then, eventually, like weakness.

Bush's seemingly unbeatable approval numbers slipped, and Clinton stormed into office having proclaimed his belief that the United States could do something about the horrors in the Balkans.

Now, as the fresh first post–Cold War president at the helm of what was now the world's sole superpower, Clinton watched the news from the Balkans getting continually worse. He read of a new leader who'd stepped onstage to speak on behalf of the Serbs. Slobodan Milošević, a boxy-looking man who had the bearing of a polite, curious accountant, but he gave a name and a brand to Serb dominance.

To Clinton, Slobodan Milošević and the Serbs looked like the guiltiest of all the fighting parties—the ones not just trying to defeat enemies but eliminate them—and Clinton thought he had an answer.

He'd believed since the footage he'd seen of El Dorado Canyon in '86, and then in Desert Storm in '91, that smart bombs could be a silver bullet.

As activists and journalists began drawing more attention to Sarajevo, Clinton listened, with a growing conviction that precision air strikes could be used to solve the spiraling crisis.

His main hesitation was that he didn't want to launch a military operation alone. He thought it better to lead a coalition of allied countries in a campaign of precision air strikes, but for months he tried and failed to secure the support of allies.

Only when Serb forces took over a weapon depot guarded by United Nations staff—a blatant affront to the UN and its member countries—did the allies agree to join Clinton's air strikes, and even then just to a small, two-day campaign.

And when the Serbs responded by rounding up UN workers and chaining them to potential targets, not all Western leaders saw it as a declaration of war against the West. Some considered it to be proof that military force was counterproductive. Allied support for more decisive action waned.

By 1995, Clinton decided he was done with patience. His reelection campaign was ramping up; soon political pressure would weigh more heavily on every decision he made. He wanted a second air campaign, and he had to act soon.

Clinton called in Tony Lake, his national security advisor, and tasked him with developing a new, more muscular strategy to resolve the Balkan crisis once and for all. Lake developed a simple, elegant plan that used Clinton's favored tool: first, a barrage of precision air strikes against Serb military positions to drive Milošević to the negotiating table. Then, if he failed to negotiate in good faith, the strikes would resume. If the other parties—the Bosnian Muslims and Croats under siege—failed to negotiate in good faith, air strikes would cease altogether, and Slobodan Milošević's opponents would be left to defend themselves.

Clinton liked the plan and dispatched Lake to Europe to try to convince the allies, with just one adjustment: Clinton said to tell the allies he was going to proceed even if they weren't. He preferred not to go it alone. But now he was willing to.

While Lake began his tour in Europe, finding NATO leaders impressed by Clinton's resolve, events on the ground conspired to make America's pitch even more convincing. Milošević launched his own endgame plan, seeking to finally, *fully* suppress his opponents by that winter. He launched a series of attacks against Muslim enclaves, and in one Muslim town, called Srebrenica, his forces carried out a mass execution, killing as many as 8,000 Muslim men and boys in the span of five days. It was an operation on a scale and carried out with an industrial detachment not seen since World War II. Images from the massacre were seen all around the world, and many Western officials considered it not just a crime against humanity but a demonstration of impunity: letting Milošević go unpunished would signal permission to other genocidal leaders. And with images from the Srebrenica massacre still fresh, news of more horrors emerged. In Sarajevo, a Serb mortar shell landed at a crowded marketplace, killing dozens of civilians and injuring almost a hundred.

NATO leaders fell into place. Two days after the market explosion, sixty NATO aircraft carrying hundreds of Paveways took off for a second air campaign in the Balkans, this one a series of coordinated attacks against Serb air defenses, weapon caches, and communication facilities. Almost 70 percent of the bombs dropped were Paveways, making it the first air campaign to depend more on guided bombs than unguided ones, and allowing the operation to proceed with such "one target, one bomb" economy that NATO commanders faced an almost unfathomable problem: that the bombing might be *too* successful. So efficient, they worried they wouldn't get Milošević to the negotiating table before they ran out of targets.

Just before they did, after eleven days of air strikes, Milošević agreed. The other key parties were more easily convinced, and soon the Balkan leaders were all departing for America, where Clinton's advisors sought a new kind of venue for negotiations. They wanted someplace different from the gilded halls where previous treaty negotiations tended to be held. Someplace more stark, and where the United States would have a nearly military level of control. Someplace accessible to leaders in D.C. but isolated enough that they could keep the parties from negotiating through the media. A military base made sense—an Air Force base more fitting still—and someplace in the Midwest would be the right mix of remote and accessible. The Air Force base that best fit that bill was Wright-Patterson Air Force Base, just outside Dayton, Ohio. Which meant the agreement, the "Dayton Accords," would be negotiated in the same place where, thirty years before, Weldon Word first pitched the Air Force brass on the very weapon that had driven them all there. And with that weapon Clinton had gotten his way: precision air strikes as a tool of diplomacy; Paveway as police force restoring order in a new unipolar world; Paveway providing a chance to lower the temperature, to stop the killing, and, especially, to force the Serbs to lift the siege of Sarajevo.

5.

Tyler, Texas

As a new method of policing debuted to a confusing post–Cold War world, Weldon Word watched from retirement as confusion unspooled through the defense industry too.

He was trying to stay more toward the sidelines now, driving in from Tyler whenever a TI colleague invited him to a wedding or funeral; otherwise he was inclined to stay out of the way and let his disciples carry Paveway and precision warfare into the future. But it was proving more difficult than expected. He'd started getting calls from journalists and academics and no longer had corporate PR to run interference or help him redirect press interest to others.

He responded to inquiries by reminding people he was retired. It didn't seem to lessen their interest, and eventually he started calling old TI colleagues and asking them to step in. As much as he wanted to move into a quieter post-TI phase of life, it was proving impossible to be entirely removed from the company and, therefore, impossible to be entirely removed from the challenges the company was beginning to face, as a compass that guided the development of weapons for his entire career disappeared almost overnight.

With the collapse of the Soviet Union, a reliable adversary was gone, and uncertainty was leaking into assembly lines. Politics were shifting unhelpfully. A common logic among triumphant politicians held that the West had won the war by simply out-purchasing the East; that the Soviet Union had bankrupted itself trying to keep up. Now that the war was over, it was time for the winners to claim a "peace dividend" from all that investment. Politicians likened this new era to the one after World War II, when factories churning out tanks and bombers returned to building sedans for smiling families, and an America that no longer had to buy thousands of tons of military hardware for a foreign war could instead build thousands of miles of interstate highways for itself.

Weldon's former colleagues now had to plot a course forward as military spending declined. Bases closed, aircraft carriers were retired, orders for munitions shrank. As demand contracted, the industry began to consolidate, and defense companies that had thrived during the war tried to survive the peace by acquiring each other and merging. Lockheed Corporation joined with Martin Marietta to become Lockheed Martin; Northrop Corporation acquired Grumman Aerospace and became Northrop Grumman. A smaller community of companies began fighting for the few remaining slices of a shrinking pie.

Defense executives desperate to prevent investor panic looked for products that a more frugal American military might still spend on. They figured the United States no longer had to worry about stocking up for a major conflict with a superpower, so if the Pentagon was going to spend on anything, it was most likely to be on weapons suited for smaller conflicts and perhaps electronics to upgrade existing platforms. That, and innovation, since even in peacetime an edge against future threats was valuable. So, in the scrum of a reorganizing defense industry, two product categories emerged as increasingly attractive to companies trying to survive the contraction: small weapons and advanced technology. There were few companies that excelled in either category. Texas Instruments happened to be one that excelled at both.

For those who'd developed Paveway and the other weapons that further established TI as a leader in precision warfare, it began to look like they'd been building what was now an almost ideal acquisition target for a defense behemoth looking for contraction-proof defense products.

Bids began to come in. In the end, it came down to Northrup Grumman making one of the most forceful pushes for Paveway, but a

Massachusetts-based company called Raytheon ultimately emerged with a winning all-cash bid. For just under $3 billion, Raytheon took ownership of Paveway and the rest of TI's defense electronics division, and the Paveway team saw the business they built—torchbearer for an industry they effectively invented—carved off from the company they'd built it with and placed under new management.

Paveway now belonged to more conservative guardians. If Weldon's Texas Instruments colleagues were no longer quite the rebels they'd once been, they'd at least still tried to channel their big risk-taking Texas oil-men forefathers. But now Paveway would be managed by East Coast accountants in Brooks Brothers suits making their sensible decisions and eliminating risk based on spreadsheets and actuarial tables, and at TI the mood among Paveway team members began to change. From confusion, as they adjusted to life under their new corporate overlords, to exasperation.

Almost immediately, Raytheon made changes. One of the first things they did with Paveway was relocate it. First just to a different city in Texas, then out of state. They'd also just acquired part of the Howard Hughes company's missile business based in Tucson, Arizona, and wanted all their missile work under one roof.

Raytheon offered relocation packages to TI employees who'd worked on Paveway, but now there were plenty of jobs in a Texas technology industry that, largely thanks to TI, was booming, and many of the Paveway team members found the prospect of a new home and new managers less appealing than the opportunities nearby. So as Raytheon began exerting its influence over Paveway, many of the people who'd been working on it for decades decided it was time to step away, to let Paveway move to Tucson without them, and to hope it would fare well under the new chaperones.

It didn't.

6.

Clinton's air campaign had succeeded in driving the parties to the negotiating table, but the goal of the Dayton Accords was peace, not pristine justice. The accords led the way to the Serbs lifting the Siege of Sarajevo, but to make sure they didn't walk away from the negotiations, the agreement didn't strip them of all the territory they'd taken—even though they'd taken some of that territory in campaigns of ethnic cleansing and mass rape. Among the territory the Dayton Accords allowed Slobodan Milošević to keep was a landlocked republic right in the middle of the Balkan Peninsula, named after a storied fourteenth-century Serb battlefield known as the Field of the Blackbirds—in Serbian, *Kosovo Polje*.

Despite Kosovo's place in Serb lore, over the centuries it had become almost entirely Muslim, so even as the Dayton Accords put an end to much of the violence, a group of Muslim rebels in Kosovo began to push back against their new Serb rulers. They called themselves the Kosovo Liberation Army, or KLA, and began to skirt the line between saboteurs and terrorists. They started attacking Serb policemen, Serb police stations, and Serbian government officials. They wanted to weaken Serb authority, but they'd also seen the Serbs lay siege to Sarajevo, and then saw the siege lifted after foreign militaries drove them to negotiate. They saw U.S.-led air strikes change Serb behavior once before. They were trying to provoke more.

The KLA steadily increased their attacks against Serb authority for three years, until in 1998 Slobodan Milošević gave up all semblance of restraint. He began attacking KLA strongholds, then he attacked perceived KLA sympathizers, and soon his Serb forces were bringing the full force of a mechanized military against families, neighbors, and whole towns with only tenuous connections to the KLA. Virtually every Muslim in Kosovo was a viable target. The attacks were at least as brutal as anything Milošević had done before, a transparent attempt to not just eliminate the KLA but cleanse Kosovo of an entire ethnicity. And it was succeeding. Hundreds of thousands of Muslims were forced to flee the country, and even more were internally displaced.

From Washington, Clinton saw violence reaching a scale you could see from space: body counts so high that satellites could identify mass graves. He'd waited in the Balkans before, letting the Siege of Sarajevo continue for almost four years, only to find air strikes were indeed what was called for all along. Here was a chance to get it right from the start. He sent his secretary of state, Madeleine Albright, to lay out his administration's position to the public.

"If we have to use force," Albright said on the *Today* show, "it's because we are America; we are the indispensable nation. We stand tall and we see further than other countries into the future, and we see the danger here to all of us." She described an emerging view of military power in the face of foreign crises, and announced America's role in a new, unipolar world. "I know that the American men and women in uniform are always prepared to sacrifice for freedom, democracy and the American way of life."

But with precision weapons, Clinton didn't think all that much sacrifice was even necessary. You didn't *need* American men and women on the ground. Paveways could get to a conflict faster than ground troops anyway. Precision weapons kept troops out of danger, and Clinton had seen Paveways be more discerning than soldiers. Precision air strikes were a humanitarian way to wage war, and he wanted his next intervention to be a war with a humanitarian soul. Refugees would be brought to safety at an American base in Cuba called Guantánamo, while America and the UK and the rest of NATO would send in planes with precision bombs used not to take over some territory for America or vanquish one of its own enemies but as a force for good, to stop a genocide. To achieve all that, it would need to be as precise as any operation before it. It would depend, more than any operation before, on Paveway.

And Paveway was in crisis.

7.

Raytheon Missile Systems
Tucson, Arizona

The few TI hands who followed Paveway to Arizona struggled with the adjustment.

They had to assemble Paveway in a tent because Raytheon hadn't built any dedicated space yet. The Raytheon culture was a shock. Engineers who'd enjoyed $5 million signature authority in Texas found themselves unable to spend a fraction of that without dispatching multiple requests through layers of bureaucracy. Even machinists ran afoul of new rules: Raytheon's workforce was unionized, so TI veterans on the assembly line kept getting rebuked for wheeling equipment onto someone else's territory or trespassing on one or another protected area of labor.

Paveway units started coming off the line riddled with glitches. Prototypes failed. Raytheon had to discard guidance units because new engineers weren't building the seekers correctly. Output collapsed, and so many Paveways were scrapped that the working units had to be sold at three times the price. The Pentagon began to worry about maintaining even the modest stockpile of Paveways it needed for a post-Soviet world, and they soon lost faith in Raytheon so completely that they asked another company to help, stripping Raytheon of its role as sole provider of Paveway II and asking Lockheed Martin to try their hand at manufacturing it as well.

Raytheon was losing face. Engineers started calling the members of

Weldon's team who stayed behind and asking them to help troubleshoot over the phone.

To members of Weldon's team still in Texas, it started to feel like Raytheon simply didn't understand the culture around Paveway. They didn't understand its *personality*, and one of Weldon's protégés found a clue to Raytheon's Paveway problems in an issue he'd heard about with another guided bomb Raytheon had taken over from Texas Instruments. The Joint Standoff Weapon, or JSOW, had started missing targets so badly that the military investigated and then started disciplining pilots because they weren't just missing their targets at a high rate; when they dropped multiple JSOWs, the weapons tended to miss by the exact same distance. It looked like the pilots were doing it on purpose.

Weldon's protégé asked the Raytheon engineer on the phone if they'd fiddled with the weapon's software at all, and the Raytheon engineer confirmed that, yes, in fact they had, but only to improve it. "The TI guys screwed up the wind estimation algorithm," he said, "so we corrected it." He said they'd run the weapon through thousands of computer-simulated drops and adjusted the guidance algorithm accordingly.

Weldon's protégé looked into the computer simulation and, after some digging, found the problem: Raytheon's simulated drops had considered wind but not the fact that wind is different at different altitudes. Raytheon optimized JSOW for a world in which wind speed five feet off the ground was the same as wind speed five miles off the ground. They'd optimized it for a world that didn't exist.

In Texas, Weldon's old team started looking into other "improvements" Raytheon may have made to Paveway and, sure enough, they found one. A Raytheon engineer said he'd fixed one of Paveway's shortcomings: the fact that it had blurry vision. He said they'd upgraded lenses in the seeker head in order to, in effect, put prescription glasses on nearsighted eyes and to focus stronger, clearer light into Paveway's brain. But what Weldon's team had understood, and the Raytheon engineers did not, was that Paveway's eyes were weak by design. Its "brain" didn't need that much information; it only needed to know whether the light was on one side or the other. Not needing an especially sensitive sensor had helped Weldon keep Paveway's cost so low. Too much light saturated it, so Raytheon's design change was less like putting glasses on nearsighted eyes and more like blasting them with a blinding searchlight.

Gradually, one by one, with help from Texas, Raytheon eliminated the problems that had begun to plague Paveway. Eventually they started building it with a consistency approaching what it had once been in Texas, and soon, with Lockheed Martin also manufacturing Paveway, the weapon was rolling off assembly lines in large numbers at two different companies.

Arsenals began to fill back up with Paveways.

8.

Oval Office
The White House

On March 24, 1999, Clinton sat behind the *Resolute* Desk to give a prime-time address announcing the start of Operation Allied Force. He characterized what the Serbs were doing in Kosovo as "not war in the traditional sense" but rather "an attack by tanks and artillery on a largely defenseless people." He'd seen Milošević promise to lay down arms once before, after the Siege of Sarajevo, only to pick them back up. So this time, Clinton said, he'd be denied the chance. NATO would use air strikes to "seriously damage the Serbian military capacity" because "if President Milošević will not make peace, we will limit his ability to make war."

That meant not just attacking the Serb troops actively engaged in hostilities but also destroying their weapons housed elsewhere in the country. It meant sending guided bombs to strike the places militias trained and the places missiles were maintained. It meant destroying the infrastructure that delivered fuel and lubricant necessary to sustain a mechanized army, attacking factories that made weapons and the institutes that developed new ones, as well as the places they were stored. That included one arsenal in particular: a cache of weapons that had been tracked by a spy satellite as it was removed from a facility in a place called Mostar and transported 120 miles inland, where it was unloaded at a factory in a remote and otherwise anonymous town called Lučani.

9.

Lučani, Serbia

By four years old, Kiki was a colt, running down the streets with the neighbors' kids, disappearing for hours at a time. Venturing farther and farther from home, finding small worlds on the outskirts of town. Neighbors' homes, little stretches of forest, and, most alluring of all, the alleys in the shadows of the local factory where all her friends had uncles or aunts or parents working. The factory was the thing, like a local football club or a coal mine that swallowed half a town's population at every shift, that worked its way into all conversations. A thing everyone talked about, because life here was small.

People knew each other. The grocer a close friend; the lady at the video store something like an aunt. Kiki dragged her mother there every week to rent the same movie and a machine to play it on: *White Goat*, about a white goat called White Goat, still somehow mysterious to Kiki every time she watched it. Re-renting it so many times that the video store lady finally told her to just keep it, and its theme music became the soundtrack to life in the family's small apartment, so Kiki began waking up to her mother singing its jingle every morning.

For Kiki, it was a day of exaltation when the video store lady suggested a new movie, and Kiki wore the mylar off that one too: an animated adventure in a place called Africa that astounded her. She'd never seen a human being who looked so different from the people in her little factory town. Her universe was Lučani, and it had always been bigger than her imagination anyway. The new movie blew the roof off her world; oxygen rushed in and fired her imagination. Even at four, at five, Kiki was developing a wanderlust, the kind that arises when you know there's spectacle outside and you know a home to come back to is always there waiting. Someday, somehow, she felt already, her little pilgrimages would take her farther, past the factory's hidden lots and corner nooks and out into one of those other worlds. She lost herself imagining those places, but never in the reverie did she picture people in those places outside thinking about Lučani.

And then one day Kiki heard a roar in the sky—a loud and slow machine wail, like the sky was tearing apart so a demon could scream at her.

"Don't worry," her mother said, smiling. "It's nothing to be afraid of. It's just—it's the new dinner bell!" Her mother's face now a frown, mock-serious. "So when you hear that, you hurry home for dinner!"

An exciting new system, then, a dinner bell so loud she could hear it all over town. Her mother, perhaps excited by this new system as well, began using it all the time and Kiki learned to save room by moving food around her plate because there were sometimes three dinners a day.

Then one night Kiki was dreaming of the jingle from *White Goat*, and suddenly she wasn't dreaming anymore. Her mother was singing in her ear, but it was still so dark that it couldn't be morning, and the apartment shook as if laughing, her clumsy home fumbling a window onto the floor. And now they were playing a game: her mother scooping her up and running somewhere, Kiki riding in her mother's arms down into a basement room near the big laundry machines. Kiki blinked, and now they were packed into a tiny car they shouldn't have all fit in, a strange vacation her mother had forgotten to tell her about. There must have been some bus they were trying to catch, some kind of deadline, because everyone seemed to be rushing. Riding through the night out of the town, through the countryside, all confusing and exciting and wonderful.

And then she woke to birds chirping and the sun shining, the sound of other children already playing outside. She was in some strange and wonderful home that belonged to an uncle, and now her mother was sitting at the foot of her bed again, singing the song from the movie again, "*Wake up, wake up . . .*"

10.

It was a warm mid-spring day, Radica on her way to work, when she understood she'd been trying to keep the war at bay by power of imagination and no longer could.

The air raid siren sounded, but the air raid siren had been sounding so often, she'd learned to dismiss it. Only this time people began to walk a little faster than her. They bumped her as they passed.

Then they were running by her, and there was a shift in her, from thinking they were foolish to thinking maybe *she* was. A current was leaving her behind. Panic seeped in and an armor of denial fell off her like a dropped curtain.

Now, suddenly, she could see the strained faces of terrified neighbors, the tension rippling in the roped muscles of their necks as they ran. Fear passed as contagion. Suddenly everyone was sick with it.

Then she was running, too, and thousand-pound bombs were getting closer, broken glass spraying across the street splayed reflected light up from her ankles and horrified her, something about light coming from the wrong place. The explosions got so close, they felt like detonations cracking inside her skull, clogging thoughts before they could finish forming; it became hard to remember where she was going, what she was supposed to do. Where was Kiki? Stores she'd shopped at yesterday spilled their windows out into the street; bodies were ejected off sidewalks. Radica was jumping over them and she knew she was next.

In a haze she reported to work even though work was close to the factory that everybody had a family member working at, the factory shrinking as flying things from the sky took chunks from it, and she felt it was just a matter of time before the explosions that kept getting closer and closer would get too close.

Inside the radio station her boss ran to her with a hammer and a socket wrench. *What am I supposed to do with these?* Her boss was a caring man but perhaps confused. He said, *If a bomb hits the building . . .*

She still didn't follow. She could see he'd had to explain this to others. He tried again: *If you get buried. You bang these against the rubble.* He shrugged. *Maybe someone will hear you and come help.*

The next day, more bombings, and she came back to work again, she had a duty here, and a duty to her family, too, but home was farther from the factory, and Kiki knew to stop her wandering, so maybe Kiki was safe.

The day after that, her boss said there was a new part of her job. This was no longer the time to do a call-in radio show for young mothers. That was not important now, he said. Now the job was, when she heard planes overhead or the air raid siren, she had to alert everyone over the radio. So those in cars who might not hear sounds outside over their engines, and people hard of hearing leaning in to the radio, and those hiding in their underground laundry rooms who *thought* they might have heard the early wind shifts of an incoming air raid might know for sure.

That night at home there was an explosion so close, it shattered an apartment window, and she knew being a little farther from the factory wasn't enough protection. She was suddenly worried for more than just her daughter's mind; she worried now for the child's life.

She woke Kiki and rushed her underground to a makeshift bomb shelter, waited for a pause in the bombing, and arranged for a car to get them out of town into the countryside, somewhere far away from the factory that attracted so much of the bombing. To an uncle's house, where other families were starting to gather. It was half an hour outside of town, far from the factory, or far enough that they couldn't hear the air raid sirens. Her husband stayed with them there, helping to keep the children busy, building up a whole ruse that this was all just a vacation, and together they made sure Kiki didn't catch even a glimpse of the truth—not while it was happening, not for years, not until Kiki was old enough to explore her history on her own.

Because even now, even haunted by the sight of broken buildings and the bodies of her neighbors back in town, Radica felt a draw. There was a war order, people who could work had to work, but it wasn't just that. Something hooked her rib cage and pulled her back into town even as bombs were falling. She had an obligation. To go back to her job. To clutch her tools and warn townspeople over the radio. So after a few days in the countryside, she began leaving Kiki there every day, taking a bus back to the radio station next to the factory.

Her boss wired up a new makeshift studio in the basement. Radica put the headset on, pulled the microphone close, and tried to ignore the sound of her own voice shaking in the headphones.

She played music for whomever was listening: American music, Serbian music, rock and roll, pop. Then the music shut off automatically in mid-song: dead air in the middle of an unresolved note. She looked up at her producer. He held out his hand and gave her a thumbs-down, their sign. She nodded.

"Go and hide," she said into the microphone, to her family, her friends, to all of Lučani. "Go and hide in safe places."

Now she could hear it, too, even from the basement. "People, sirens are on." She pictured resistance. She imagined people trying to make the American bombs disappear by ignoring them, as she had. She imagined them simply not hearing her.

"I *need* all of you to go to the safe places," she said. "Until the sirens stop and you can come out. And don't worry, we're going to tell you when you can come out."

She took off the headphones and went to hide under the desk, next to their little gas camping stove and her supply of water, so even if the building came down around her, she might still enjoy her mud-thick strong coffee for a day or two before she died or someone came to rescue her.

She crouched down there and wondered what gave NATO the right to terrorize her people like this. What gave Americans the confidence they could fly over her country and know what to bomb, to think they knew anything about her country, to think they knew what was really happening? A visitor barging into a home and acting like they knew your family better than you did. And why? Her people were welcoming and hospitable to everyone. It'd always been the thing she was most proud of. The instinct Serbs had to welcome others, even non-Serbs, *especially* non-Serbs. Maybe the cruelest thing you could do to a welcoming people, to a people who'd invite you in, was invade.

Every day she returned to Kiki out in the countryside, then came back into town to work, braving the raids, tempting the chance she'd be leaving her daughter motherless, because she had to warn all these people she knew and many she just imagined.

But you can't talk back to the radio, and everyone's phone was down. No one ever called in to say they'd made it to safety. She never knew, when she warned people, if they survived. She was speaking to her whole extended family, but she was also speaking to a void. Speaking to all these people who might be gone forever without saying goodbye.

11.

On June 3, 1999, after more than two months of bombing, Slobodan Milošević capitulated.

His militias were overwhelmed by the assault of precision bombs. Paveways had devastated the power grid and damaged many "dual-use" targets too: oil and petroleum facilities used to support the military and civilian infrastructure. Factories had been destroyed, and Milošević faced the prospect of his people turning against him, as any more bombing would mean a harsh winter without the benefits of consistent power and heat. He accepted a peace proposal mandating that he withdraw all troops from Kosovo and give up Serb control of the region.

It was "a victory for a safer world," President Clinton said, in another nationally televised prime-time speech from the Oval Office. "Aggression against an innocent people has been contained, and is being turned back."

Once more, precision airpower had won the day. In Libya, El Dorado Canyon had proven you could use Paveway to strike faraway targets without major military mobilization. In Iraq, Desert Storm proved it could defeat a dictator. Now, in Kosovo, Operation Allied Force showed that Paveway could not only defeat a dictator; it could do so virtually by itself. Airpower advocates claimed Kosovo as evidence that now precision airpower alone could win wars.

For Clinton, the lesson of precision airpower was grander still. "We have sent a message of determination and hope to all the world," he said, to close out his address. "Because of our resolve, the 20th century is ending not with helpless indignation but with a hopeful affirmation of human dignity for the 21st century."

12.

Lučani, Serbia

Things weren't quite right in Lučani.

Radica could see, even after the bombing stopped, that the violence was just beginning.

Her family had escaped the bombing mostly unscathed, and it didn't matter to her that the Serbs had lost. Whatever "Sloba" had been up to, it was so remote to her, it wouldn't have figured into her life at all had he not invited the might of the world down on her little town.

She'd managed to keep Kiki in the dark—or, really, her husband had. And when the skies went quiet, she brought Kiki back from the countryside and they tried for normalcy.

Radica's father tried gardening. But when it came time to pick vegetables, what he pulled out of the ground sickened her. Yellowed, necrotic things. She was sure the American bombs had somehow done this. They'd killed her friends, poisoned the minds of children, and now she was convinced they'd poisoned the ground too. Around her, people began to die of strange illnesses: too-young people having heart attacks, strange new cancers.

And when she became pregnant with a second child, it wasn't a reprieve, it didn't feel like a blessing, it felt like running toward a cliff's edge. She carried some deformed thing inside her, she was sure of it; she knew what grew inside her was some festering goblin from what these bombs had done to the ground and maybe the air. She needed to leave.

They decided to leave Lučani for the big city, for Belgrade, where her husband used his engineering degree to get a decent job. There was plenty to rebuild.

But Kiki hated the move. She hated Belgrade; she cried all the time. There in the big city she couldn't run free with her friends like she used to in Lučani. In Belgrade she didn't *have* friends. Her one comfort was watching TV with her grandmother, the two of them bonding over a saucy Mexican soap opera that, by some accident of global commerce, was beamed into the middle of Serbia. *Rebelde* was about rebellious adolescents double-crossing each other at a fancy prep school in Mexico, so Kiki's only friends were rich kids double-crossing each other in Spanish with Cyrillic subtitles.

When Radica gave birth to another healthy daughter, it felt like she and her husband had narrowly missed a catastrophe. They began to think they should be someplace safer still, someplace better, new, maybe very different, with healthcare and space for Kiki to run around, and they must've discussed it out loud, because one day a relative said he'd submitted their names to a visa lottery he'd heard about for, of all places, America. Was that the only option? Running to the country that had sent the bombs that destroyed her home and poisoned the ground and maybe the air?

And in a second stroke of luck, or maybe it was a curse, they somehow won.

Could they really move to a place whose planes they had to run from? None of them spoke the language. The closest any of them came to speaking English were prep school insults they could halfway parrot in Spanish. The idea of America was dense and impenetrable.

But there was nothing for them there in Serbia, and if America was the place that sent the bombs, it was at least a place where bombs never seemed to land. Her husband had a distant uncle in a place called Florida, so they'd only be mostly alone, and they figured: Why not take a leap? They packed up their small apartment, said their goodbyes, and the family ventured west. A cliff dive into a new life that would lead Kiki to a kind of collaboration with the people who bombed their town and set Radica on a journey to a new calling, to communion with unlikely allies.

And that would lead her, eventually, to me.

Chicago, Illinois
United States of America

Kathy Kelly sits at her father's bedside. She thinks she doesn't have to feel what she's feeling.

She's in touch with this superpower she has, to disassociate, honed in a handful of war zones now.

She can sit there with a kind smile and place her mind somewhere else. She can trigger it almost at will.

But she doesn't. This is family. And she's been gone so much.

When her father began his long, slow decline, she skipped another trip to the Balkans and moved him into her Chicago apartment. She feels—she knows—this is not something everyone gets.

She's been around for the obscene instants when lives in other places tend to end, so she chooses to be there with her father now. She knows she should feel lucky. It's a grim privilege, helping a parent tread water over deeper and deeper swells of depression.

She's there for chemistry's coordinated assault on his body and mind, accelerating cell death, gathering until it's large enough to see. He sinks into his bed and withers.

She tries to be present for him. There is a planet of people she isn't helping. She won't be traveling to war zones with any Peace Team members anytime soon.

But soon, it turns out, they begin traveling to her.

Old friends come by Kathy's apartment to talk, even though her father is always there, a proud envoy of the Greatest Generation. A high school teacher who enlisted so he could go fight Hitler, now trying to understand the radical his daughter has become. This business of getting in the way of the greatest country fighting the greatest evils, and he doesn't mind saying so. When he can muster the energy, he jokes about the growing coterie of radicals at his bedside. They come anyway.

And then they come because of him.

Social calls turn into something more formal. War stories zip back and forth over Mr. Kelly's chest, and plans are made before him. Delegations formed to go break sanctions, to stand in refugee camps and get in the way of military operations, to protest nuclear arsenals. Before Kathy quite understands what's happening, her cozy Chicago three-bedroom apartment has become a Great Lakes beachhead for her squad of dedicated activists, a sleeper cell coalescing around its centerpiece, a dying veteran.

A ritual forms. Peace Team members arrive, greet Kathy's thinning father, ask if he needs anything, squeeze his hand, and then get to the planning. Mr. Kelly became a kind of muse, trapped in his body and ensconced in a thickening depression, but breaching the surface every once in a while to smile and throw jabs.

He'd once huddled in tunnels under London while Hitler sent vengeance weapons to kill as many civilians as he could. Did that not justify violence to stop? Not even death camps?

"Well," Kathy says, "I just wonder what would have happened if the peace after World War I hadn't been so punitive to the Germans. If they'd been allowed some dignity, rather than making Germany such a poor, desperate country, would Hitler even have been possible?"

An answer for everything! But, still, Mr. Kelly seems to like the arguments. He seems to relish the role of practical old curmudgeon, tsking at the blind and deaf next generation. *How do you even pay for these antics of yours?*

They have an answer for that too. *Because we're stealing your Social Security!*

He likes the company, and these friends of his daughter's have become his friends too. And maybe Kathy isn't out at the front lines as much as she might otherwise be, but she decides this is still a fine way to spend the end of the millennium. Watching her father's unexpected friendship with her anti-war buddies.

The year 2000 comes, and she sees him pass on peacefully enough.

The year 2001 arrives, and if Kathy and the team members see plenty of injustice, outside in political America, there are also signals that the world might be on a path toward becoming peaceful enough too. Chaos in the aftermath of a collapsing Soviet Union seems to be easing, with countries and leaders receding into their new roles. Clinton is leaving, and with him that proud brand of martial impulse. George W. Bush comes into office sounding a note of caution, again and again, about war. Campaigning on the notion that America's military is overextended, that now should be a "time of pride and promise." Campaign speeches saying that involvement in the world "does not mean our military is the answer to every difficult foreign policy situation." He says he won't do what other presidents have done: react to the world, let their agenda be set "by adversaries, or by the crisis of the moment." He pledges that no matter what the future holds, he won't go looking for foreign military adventures, and Kathy watches 2001 dawn with at least some reason to hope for a less violent America, a less violent world.

BOOK VI

LUIS

(SHOCK AND AWE)

1.

Central Intelligence Agency HQ
Langley, Virginia

Luis Rueda was finding he wasn't at all prepared to be deskbound.

He'd spent the bulk of his CIA career thus far in covert action, running ops in the Agency's Latin America division. The Latin America division was a good fit: behind enemy lines, or at least all mixed up in them. Armed and in the middle of the action, moving around like a cowboy. Running strike teams, organizing paramilitary ops, leading buses full of trigger pullers, feeding left-wing journalists right-wing propaganda. Recruiting spies, turning people, vetting sources, running the polygraphs, getting better weapons to the good-ish guys in the jungles, and eventually managing other case officers the Latin America division sent down to him. A few tours feeling he was running an entire intelligence agency in miniature. He'd done a tour there and thrived, was sent down for a second, then a third, a fourth, it was unheard-of to do five—he did a fifth, and by then he'd finally started to sour on the work. The CIA's chief adversary there shifted from communist insurgents to drug cartels, so Rueda's brief shifted from fighting his childhood nemesis to fighting with other American agencies for who could do the shittiest job keeping cocaine from leaking north. It was gritty, violent work, sometimes extravagantly violent, and he'd been around enough death that he could feel overexposure corroding his soul.

But now, back in Langley, he was finding he'd been able to play the overlapping loyalties of a guerrilla war better than he could navigate the hair-trigger egos filling up seven floors of Washington bureaucracy. He cursed, he drank quadruple shots of espresso, he butted heads with people. His drive for career advancement weakened, disappeared, and finally gave way to a desperate need to do something. He was bored. He needed a rescue, a lifeline back to some deadly fucked-up war zone of a place, anything to break up the monotony. And when George W. Bush took office in January 2001, Rueda's prospects for excitement only dimmed. The home-focused president who campaigned on restraint, a boring foil to Clinton's vision of America as the world's policeman.

So it was strange to Rueda when, a little over a year into his tour as executive assistant to Deputy Director John McLaughlin, he started to see announcements for meetings about "Persian Gulf affairs" two times a week, sometimes three.

What the hell was "Persian Gulf affairs"? Bush was supposed to be pulling back, and if there was any pet rock to explore, it'd make more sense for it to be Russian, since his national security advisor was Condoleezza Rice, and Condi Rice was well known as a Russia expert. Stranger still that the agendas for the "Persian Gulf Affairs" meetings arrived at the last minute, and the minutes were rarely circulated. It seemed the people around Bush were trying to keep it close hold. And maybe it was wishful thinking, but Rueda figured it could mean that despite the campaign rhetoric, they were gearing up for some new military action. Wide-eyed for an opportunity to do almost anything, he finally asked his boss what "Persian Gulf affairs" meant.

"When they say 'Persian Gulf affairs,'" McLaughlin said, "they're talking about Iraq."

IT WAS A surprise, but less of a surprise once he thought it through. If you were looking for it, the signs had been there even before Bush took office. Desert Storm had driven Saddam out of Kuwait and depleted Iraq's defenses, but the dictator had held on to power. A series of secret plots in the '90s failed to oust him, and by 1997 future Bush officials were no longer bothering to hide their opinions. They began publishing journal articles meditating on ways to topple Saddam's regime, and in 1998 they wrote a letter to President Clinton urging him "to turn your Administration's attention to implementing a strategy for removing Saddam's regime from power." They produced a blueprint for the future of American defense, which argued that toppling Saddam should be a top priority. "From the Gulf War to Operation Allied Force over Kosovo," they said, "the increasing sophistication of American air power—with its stealth aircraft; precision-guided munitions; all-weather and all-hours capabilities," proved the United States could now "attack virtually any target on earth with great accuracy and virtual impunity. American air power," they said, "has become a metaphor for as well as the literal manifestation of American military preeminence." The blueprint was completed in September of 2000, two months before the election that brought Bush to power.

Now its authors were settling into senior administration roles just across the river from Rueda. And though Rueda hadn't tracked the backstory all that closely, he was primed to see—busy as he was dreaming of an escape back into the field—that Bush's closest advisors seemed to be gunning for Iraq. If there was going to be some kind of operation in Iraq, Rueda knew the CIA would be asked to support it, and someone at the Agency would have to run an operations group.

And that meant Rueda might just have a way out.

WHEN RUEDA STARTED his new gig as chief of the Iraq Operations Group in August of 2001, he found he'd inherited an empty, smoking husk of a program. If the Bush administration was serious—if the United States really was going to launch a bombing campaign there—he wouldn't even be able to tell them where to aim. He made the rounds at Langley, looking to other CIA minds for advice on how to get a covert operation up and running in Iraq. "Look," his predecessor said. "The job here is to make sure the words 'CIA,' 'Iraq,' and 'fiasco' don't appear together on the front page of the *Washington Post*."

The last time the CIA had tried to run a covert action in Iraq, a fiasco was exactly what happened. The Agency ran an op in '96 out of Amman Station and a base in northern Iraq where they convinced the Kurds—an ethnic minority and bitter rivals of Saddam—to take the risk of helping them remove the Iraqi leader. It turned out to be a risk not worth taking. Saddam's secret service infiltrated the plot so thoroughly that when Saddam struck—which he did only after waiting, listening, and uncovering virtually every part of the operation—he got nearly everyone. He had every single CIA asset rounded up and tortured, or killed, or both. The CIA fled Iraq with its tail between its legs and about a hundred dead Kurds on its conscience. Many of the officers involved were unwilling even to talk about it. The Agency had been purified of institutional memory in a post-traumatic blaze, the Iraq program was unofficially deactivated, and the entire scope of what remained amounted to little more than just updating lists of phone numbers so that if by some miracle a revolution in Iraq happened on its own, the CIA would know who to call. A week into the job, Rueda understood he hadn't taken over an intelligence operation. He'd taken over the Yellow Pages.

He looked everywhere for guidance. He devoured books on Iraq. He abused his seventh-floor status, slipped into George Tenet's office to call around the Agency from the director's desk, ordering intelligence products on Iraq. He crammed. He studied Saddam Hussein. He began to see Saddam had more than raw terror to use as tools. There was method to the madness. Saddam had a Special Republican Guard unit spying on the rest of the Republican Guard, which was in turn spying on the army while a separate intelligence service drifted among them all like a free agent: a system to not just encourage loyalty, but to guarantee it. Saddam had body doubles; he kept his daily travel plans secret even from people in his own security detail. He sometimes moved in huge armored convoys and sometimes rode in a single inconspicuous vehicle painted yellow to look like a taxi. Rueda was intimidated. He was invigorated.

He made the rounds among Bush's people to look them in the eye and gauge for himself whether they understood what covert action in

Iraq would really look like. He needed to know: Are these guys serious about invading? If he could somehow turn people and get assets in place, would we end up abandoning them to face their fates alone? Because he'd seen that movie before. In a way not everyone around him understood, it was part of his own origin story.

"We've never faced an adversary like this," he said to Bush's team. "Once you start this covert action program, and once we start things in gear, the clock starts to tick . . ."

In the White House, Bush's advisors tilted their heads. They said they understood: *This is real. What's the next item on the agenda?*

Rueda wasn't satisfied; he felt something bordering on compulsion. ". . . And by that I mean we get closer to getting wrapped up. We can go out there to try and recruit people, and you pay them money, and maybe, *maybe*—slim chance—we'll help them get out of the country if something goes wrong. But you're facing a system where if they get caught, they're tortured to death; their families are tortured to death. And many times the village where they live is destroyed. Everybody's killed."

We get it, they said. *We're serious. Don't worry, an Iraq invasion is going to happen.*

Rueda couldn't be fully satisfied, but setting policy wasn't his province, and if Bush's people were telling him to support a war, that's what he'd have to do.

And if that's what he had to do, then he was going to have to turn the husk of a division into just about the best group in the Agency's history. This was the ultimate "denied-area" operation, and he'd need the very best people in the field. He figured he'd need to stand up an operation of about three hundred people starting from nearly nothing, which meant applying the best lessons from his five tours recruiting assets to the task of recruiting within his own agency.

But just as he got started, half the talent pool disappeared.

Rueda was a month into his new job as chief of the Iraq Operations Group when the 9/11 attacks made Iraq seem like a foolish distraction. Officers all along the hierarchy, right up to Tenet, shouted Rueda out of their offices. None of them believed the White House could still be serious about Iraq—not after 9/11. Now counterterrorism was the focus, an all-hands priority. At Langley, the Counterterrorism Center grew by 400 percent in a matter of weeks, sucking up resources from all over the Agency and absorbing so many officers that overseas some CIA stations had to shut down. Intra-agency fights bubbled up. Egos got involved. Cooperation began to cool, and it was into these bureaucratic frontier towns that Rueda rode up, asking division chiefs for their best and brightest.

"*Iraq?* Why the hell are you asking me about *Iraq*?" Rueda might as well have been planning covert ops on Mars. Doors slammed, phone handsets were slapped back onto their cradles. Out in the Langley parking lot after an exasperating day trying to staff his Iraq operation, Rueda confided in Deputy Director McLaughlin. "It's like we're a chicken bone caught in the Agency's throat," he said, "and they're trying to hack us up."

McLaughlin wanted to help. Others on the seventh floor wanted to help. Associate Deputy Director of Operations Steve Kappes called a meeting, mandatory attendance for all chiefs of station in the Near East Division, and pounded home the message from Bush's people: "We *are* going to go to war with Iraq. We need all you guys to get on board," he said, "because this comes from the President of United States."

And the president was getting ever more eager. Rueda was summoned to the White House and asked why he couldn't just stage a coup to topple Saddam.

"With what assets?"

He was summoned by a Senate select committee and was asked, "So why don't you just assassinate Saddam?"

"You made that illegal," Rueda said.

He went before a House committee. "Precision strike? Guided bombs?"

Against what? There was still no bomb capable of gathering its own intel. How do you launch a precision strike against what you can't fucking find?

There was a sense that Saddam should be gone with the flip of a switch, but Rueda had looked at all the options. A coup wouldn't work because there were no assets to start one. A war was doomed because they had no intelligence. Precision strikes were a nonstarter because they didn't know what to hit. People started asking the same questions again, as if maybe Rueda hadn't actually thought it through. "Look," Rueda said, "I've looked at this from all sides now." He repeated the briefing so often, he began to hear it set to music, Joni Mitchell in his head: *I've looked at clouds from both sides now.* Every idea required human intelligence from people on the ground, and the only way anyone could think of to get that was from a group whose members had been tortured and killed—because the last time the CIA tried to bring down Saddam, they bungled a coup and abandoned the country. The CIA needed the Kurds, and the Kurds had every reason not to trust them.

And for Rueda there was something else. Another inkling of resistance from within him. For him, because of something in his own history, there was also going to be the matter of convincing himself.

2.

April 17, 1961
Havana, Cuba

Forty years before, a four-year-old Luis Rueda was waiting for his father to come home from work when he began to sense something was off.

It wasn't just his dad. None of the men had come home. Then men showed up, but they were men he didn't know. Standing on street corners, saying they were there to defend the revolution.

He didn't know what that meant. But the women in his family stopped talking.

His world went quiet. He could hear new things. The creaky sounds of empty streets, his mother's thumping heart. At night, the flicking whiskers of the pet bunny rabbit sleeping next to him. He was scared because the people around him were scared and he didn't know why. And because his father was gone and he didn't know why that was either. What he knew was that there seemed to be little to eat. He'd heard the term *bloqueo*—embargo, blockade—but didn't know what it meant or if the men had been holding all the food when they vanished.

He woke up in the middle of the night and his pet bunny was gone. No one had the heart to tell him what happens to pets when people are hungry.

Everything was strange; the only source of normalcy was his grandmother, who was maybe a thousand years old and had lived through all of history's outbreaks and wars and was steadfast in rejecting everyone else's concerns. She started taking Luis on her daily coffee runs, his stroller became her walker, and together they went out into the streets of Havana like it was still the den of sin and corruption and *life* and music it had been when American bankers and Mafia dons were flying down for the brothels and sometimes just for lunch. Every day, she wheeled Luis to her favorite bodega for her favorite Cuban coffee, then one day decided, *a la mierda*, and ordered an extra serving for the four-year-old. Rueda was handed his own shot: pure candy to a kid—thick, sweet cream like ice cream but hot, with an explosive dose of caffeine potent enough to launch a toddler into outer space. Months that turned Luis into the kind of caffeine addict who needed a quadruple shot just to get up to a resting heartbeat, and the kind of colleague defined by poor impulse control, ejecting streams of profanity-laced opinions before he could ever remember decorum, but so be it: this was quality time with a survivor teaching him how to find normalcy in war, another fixation that would work its way into his brain chemistry, because it's how he survived those months before he learned where his father was.

His father was alive but hiding. Luis would learn this when he finally saw him again, months later, after a journey over the ninety miles of ocean separating Cuba from a refugee camp in America.

THE FAMILY FINALLY reunited, huddled in their allotted box of flimsy nailed-up plywood at a camp outside Miami. Then his father began telling the story.

Luis's father had been on his way home from work that day in Havana when, on the other side of the island, boats full of American-trained Cuban expats landed at the Bay of Pigs. A stranger on the street stopped him. "Don't go home," she said. "All the men are being arrested."

He hid in a cemetery, then snuck to the embassy of Brazil and holed up there with a few dozen other Cuban men, and by then what had happened at the Bay of Pigs was becoming clear.

A CIA-backed brigade of Cuban exiles had landed at the bay with a mission to overthrow Fidel Castro. The CIA predicted a surprise landing would spark an uprising of support from within Cuba, but none came. The brigade was immediately pinned against the beach, outgunned and outmanned by Castro's forces, who hadn't seemed surprised at all. The brigade was trapped on the beach for a day, then two days with no food; they were running out of ammunition and taking heavy fire from Castro's forces. And though they'd been told they'd have American backup from a naval task force waiting out in the bay, in the end that backup never came. Within seventy-two hours of landing, nearly every member of the brigade had been killed, arrested, or driven back into the sea, and all over Cuba Castro's forces fanned out to find collaborators and smother any chance of a popular uprising.

Luis's father hadn't been sure if Castro's forces knew he was an activist in the anti-Castro underground, stuffing ballot boxes for Batista, raising money, and putting out pamphlets. Castro didn't seem to need a reason to round people up, so the Brazilian government negotiated a deal with Mexico, and after three months hiding in the Brazilian embassy, he and the other hiding Cubans were flown to Mexico City. When he landed, an American consular officer was waiting at the bottom of the air stairs. "Who wants to go to America?" he asked. Luis's father decided to take the man up on the offer, flew to Florida, sent for his family, and hated Kennedy maybe a little bit less for abandoning people to die on the beach.

Luis began life in America as a refugee fleeing a disastrous American intelligence operation, and—even removed from Cuba—in his family the Bay of Pigs loomed over everything. It split dinners; arguments seethed across the table. When Khrushchev parked nuclear weapons in Cuba and the world spent thirteen days wondering if humanity was about to annihilate itself, Rueda's mother and father went at each other.

"If they're not careful, Cuba's gonna go up in a fireball!"

"*Let* Cuba go up in a fireball!"

Luis watched his father. The dominant message from the senior Rueda, and therefore the one that seeped through the transplanted Rueda clan as Luis grew older, was: *Yes, the United States fucked us over at the Bay of Pigs, but they're fighting communism, and you sure as shit better do your part to help.* "If communists win the Cold War," he said, "we're on *lists*." He didn't hold back in front of the kids. "If these guys win, they'll line us up against the wall and shoot us." Luis's father made sure the whole family got the message: *It's life and death. It's real. It's war. Communists are evil, devious motherfuckers. They're a mortal threat to us. To us.*

Even as an adolescent, Luis Rueda couldn't look at the natural-born Americans in his classrooms without a twinge of resentment. *You assholes have no idea what this is like. You're safe and happy, but out there it's constant fucking war.*

America needed families like the Ruedas, and when members of the expanding Rueda clan came of age, each one followed that message right into U.S. military recruiting centers.

All of them except for Luis, who decided that if he was going to fight communists, he'd do something a little sexier. A little more fun. He wanted to do stuff from the movies. He was into spy shit. He decided: Why not try the CIA?

He had reason to be nervous at first: the Agency was terrain for America's manor born, and Rueda was an immigrant. Worse, a refugee; even worse than that, perhaps: a refugee from Cuba whose cousins on his mother's side still had links to Castro. Third-class status, and he might look compromised. He opted for an "I'm an open book" strategy, spending the day before his polygraph, filling two legal pad pages with names of family members. He went in and made a thing of handing the list to an interviewer, who put it aside without even looking at it.

The intake process went seamlessly. He passed his polygraph. He believed he may have underestimated them. Perhaps his new superiors understood that no one could be more rabidly pro-America—or at least more rabidly anti-communist—than a Cuban kid who'd fled Castro.

CIA—Training

Rueda's training in clandestine services at "the Farm"—the facility in Virginia where the CIA prepared its recruits—went well enough. Some of it was as exciting as he'd hoped; some of it was as boring as any classroom he'd struggled to stay awake through. Sundays were for Division Nights, when chiefs of regional CIA divisions sermonized to new recruits, droning on and on about why their particular corner of the world was so important. Then one Sunday the Soviet Union–Eastern Europe chief came down and blew Rueda's mind.

"The CIA exists," the chief said to the gathered recruits, "because the Soviet Union can wipe us out in twenty-four minutes." Rueda waited for the rest. "Any questions?"

In the silence that followed, it all clanked into place. He wanted to do covert operations in the Soviet Union. He wanted to do the most important work, the most difficult work, moving through the heart of a beast that didn't even need half an hour to destroy us. In the places the Agency called "denied" areas. The officers doing that kind of work weren't just ballsy; they were brilliant, they were creative, they had to be fucking meticulous. He wanted that: an outlet for his amphetamine energy. To figure out how you go about prying intelligence from some clenched fist of a city—that would be meaningful. That would be fucking fun. He'd sign up for Soviet Union–Eastern Europe, and then he'd sneak around Moscow, climb the wall, save the world, recruit people, kill people, do James Bond shit he'd seen in the movies.

Except the reality was the Agency just didn't do all that much covert work, and most of the covert work it did do was in the Southern Hemisphere. Rueda was a native Spanish speaker, which made it more or less inevitable that if he wanted to do the movie star stuff, it'd be mucking around down in the dirty wars, hoping that backing capitalists there would still be doing his part erecting defenses against the communist menace.

Now, five tours in the field, a few years at HQ, and a vanquished

Soviet menace later, he still heard the words "destroy us in twenty-four minutes" as clearly as he had the day they'd been said. This agency existed to save the country—Rueda could still say it with a sort of straight face.

But now he found, for the first time, that maybe he did have conflicted feelings. His own founding trauma was a wire wrapped around the whole-body motivation he usually had, slowing the motor.

He was still here to save his new country, or at least to serve it. But his new country was asking him to serve it in a way that something in his soul opposed. He could see his people on the beach, being rounded up and killed by a dictator because of the risk America had asked them to take. He could see them, see himself, in the Kurds who would have to risk their lives for America again. He had to find a way to make them believe that the CIA was for real this time. And he knew if he were in their shoes, he probably wouldn't. He needed to convince them to trust him. And he needed to convince himself that he was worth trusting.

3.

Central Intelligence Agency HQ
Langley, Virginia

Rueda decided he'd start by sending in money and supplies to curry goodwill. Fuck-you money, tens of millions in American currency, enough that the logistics of transporting it would be an issue. He crawled around official Washington on a charm offensive, begging for a massive allowance, then did what he could to fight off all the Agency vultures who suddenly stopped ignoring Iraq and claimed their work was related. How the fuck did Brasilia Station have anything to do with Iraq? As soon as approvals came through, he hurried down to Agency finance and had all the cash packed up before anyone could make a claim on it. Its own form of atonement: a five-figure civil servant handling a van full of cash for someone else.

He got the money, but he knew money wasn't enough. He could give his word, but he knew that wasn't enough either. He could promise that this invasion Bush's people were planning was really going to happen, but even if he believed it himself, the Kurds had little reason to take his word for it. There was really only one way he could convince the Kurds to take the risk: if he did too.

He needed to put Americans on the ground with the Kurds. He needed his own guys to put it all on the line right alongside them.

He'd need a crack team. Two crack teams, one for each of the major Kurdish parties. They'd need to be largely self-sufficient in the field. He asked for officers who'd been through the Latin America Division like him, because those were assignments you only survived if you could do it all. Logistics, leading strike teams, a deft hand at recruiting agents. He needed the best of the best.

The brass sent him a Korean linguist. *What the fuck am I going to do with a Korean linguist?* They sent him a case officer with decent experience but a pending Department of Justice indictment. It was like they were *trying* to spite him, and when he complained, they shrugged and blamed the staffing demands of an expanding Counterterrorism Center, because CTC was—at least as far as the Agency was concerned—priority one.

In the shitstorm of misfortune, though, Rueda got lucky a couple times. One guy, Tom, was an Arabic speaker and former Navy SEAL who somehow escaped the roving hordes of CTC recruiters. Rueda put Tom in charge of one of the teams, staffed him up with guys who lacked experience but who he at least had personally trained at the Farm, and gathered them all together for the world's shittiest pep talk.

"You'll never get a promotion after this. Running one of these big programs does nothing but make you enemies," he said. "Just so you know, you're signing up for career suicide."

But that was key. His team was going to be the tip of the shit spear. He didn't want careerists. Careerists wouldn't survive. Careerists hesitate. They fuck each other over. Iraq was "denied-area" operations on steroids; he needed *zealots*.

Rueda's guys grew out their beards, touched up on their language skills, and took a short refresher course on call-for-fire commands in case they found themselves in a jam. Rueda briefed them on the mission. Two simple objectives: *First, convince the Kurds we're not gonna fuck off if things get hot this time. Second, find whatever you can about Saddam because we have no intelligence and the military's got no fucking clue what to bomb without us. Sneak in, set up safe houses, start handing out cash. Chum the waters and see if you spot any sharks. Don't get killed.*

And then he sent them off to Iraq to try to build an intelligence network from scratch, and Luis Rueda began two years of more or less constant nerves.

4.

Raytheon Missile Systems
Tucson, Arizona

By the early fall of 2001, after Paveway's disastrous move from Texas Instruments to Raytheon, Raytheon had begun taking steps to improve its image, but Paveway still faced an uncertain future.

The Paveway assembly line in Arizona was up and running, but output was still below what it had been under TI, and Raytheon had lost its monopoly on the world's favorite guided bomb. Now Lockheed Martin was licensed to sell the weapon, too, and other companies, inspired by Paveway's success, were chipping away at its dominance with other guided bombs. And even as a series of U.S.-led precision air campaigns throughout the '90s further solidified enthusiasm around precision airpower, the weapon makers were under pressure from a more frugal post–Cold War customer base. The election of George W. Bush promised to weaken demand even further, given the aversion to foreign wars he'd repeated throughout his campaign.

Raytheon execs needed a competitive edge, and a close study of the military's most recent missions provided a clue as to what that edge might be.

One of the few blemishes on an otherwise widely lauded Allied Force air campaign in Serbia was the shooting down of Vega 31, one of the stealth bombers previously thought to be nearly invincible. The pilot had just dropped two Paveway IIIs, and before he left the target area, his plane was spotted, engaged, and shot down by a thirty-five-year-old missile system. Stealth technology being defeated by comparatively rudimentary air defenses worried war planners and betrayed a key vulnerability. Even though Paveway allowed pilots to bomb targets from much farther away, they still had to remain close enough to the target to keep a laser pointed at it until the bomb landed. Improvements to Paveway's aerodynamics allowed it to glide greater distances and further reduced the risk to pilots but clearly did not eliminate it. Vega 31 highlighted that there was still a problem.

Two potential solutions—two new ways of guiding weapons—gained traction. The first had to do with a weapon doing more calculations itself: if a bomb could be equipped with gyroscopes, a brain capable of complex math at speed, and the ability to determine its own angle, speed, and inertia, then theoretically it could cross-check flight data against data from the beginning of its flight and calculate whatever steering was necessary to get itself to the target. Engineers called the concept an "inertial navigation system," or INS, and while it would likely be less precise than a laser, it would be less reliant on a pilot staying nearby and less susceptible to weather than a laser was.

Meanwhile, just as Weldon had predicted, other companies were working on guidance systems that used information coming from satellites in space. Boeing had already introduced a GPS bomb, which turned out to be less accurate than Paveway but didn't require a pilot to remain in the target area, shining a laser, and was also less susceptible to weather.

And, thanks to Weldon, Raytheon already had a head of steam with GPS. Before Raytheon acquired Paveway with TI's defense electronics division, Weldon's team had done groundbreaking work on GPS. Texas Instruments had built some of the first commercial GPS receivers to make use of the growing network of satellites in space and followed the same playbook they'd used with semiconductors to make GPS receiver technology small and cheap enough to fit on a chip and therefore available to a broad range of customers. TI's GPS technology was already being used on board ships and in cars, and eventually it would be small and inexpensive enough to let mobile phone manufacturers put GPS navigation into the hands of millions of consumers. So, as Raytheon began working on a new version of Paveway that might restore the company's luster, they found that they actually already had much of the research they needed in-house.

As with every new Paveway idea, the engineering challenges would require time and money to solve, and the idea of investing significant resources to improve a product other companies sold as well seemed imprudent. But while Raytheon was no longer the sole provider of Paveway II, it did still have the sole license for Paveway III, and Paveway III's bunker buster capability was still unique. So, as the new millennium dawned, the company decided that by marrying technologies—the Paveway III bunker buster capability, laser guidance, plus INS and GPS guidance—they'd be able to offer something truly exclusive: a bunker buster bomb capable of the precision provided by laser and also the independence and imperviousness to weather allowed by GPS and INS. A bomb that could, if needed, find its way to the target even if the plane that dropped it had to leave dangerous airspace and could no longer shine a laser to guide it.

To realize that dream, Raytheon would need to give Paveway a new kind of vision because it would need to see not just a laser reflection below but a signal from a satellite above. Paveway would need to learn a new language in order to understand the signal coming from satellites, and it would need to learn something like discernment. It needed to know how to handle ambiguity and how to deal with internal conflict. If GPS told the bomb to go left and the laser was saying *Go right*, how would it resolve the argument?

If Raytheon could figure the challenges out, they'd have a flagship that might just make the industry forget about how poorly the company had handled the Paveway acquisition. But it would take time and a significant investment, and even if the promise of such a weapon seemed obvious, the military had not explicitly asked for it and refused to provide much in the way of funding to develop it, no matter how convincing a case Raytheon made.

The company wanted the weapon; the company didn't want to waste man-hours and money on a project that didn't have a guaranteed customer. Once again a new Paveway project threatened to stall.

Then, in the course of a few hours, everyone's priorities changed when, on September 11, 2001, hijackers crashed commercial airliners into the World Trade Center, the Pentagon, and a field near Shanksville, Pennsylvania, and the president's campaign promise that he wouldn't let any crises dictate his priorities was forgotten.

Raytheon saw a surge of new orders for smart bombs. The plant in Arizona had just reached the capacity to produce around two hundred Paveways per month and almost overnight needed to somehow increase production to more than 1,000. The American military wanted more Paveways; they wanted more planes capable of carrying old Paveways; they wanted Paveways on everything. Demand skyrocketed, and as America assembled a coalition for its war against Al Qaeda, more than thirty countries were soon asking for Paveway too. The Raytheon plant became a roiling hub of activity, with employees on three shifts working on assembly lines humming around the clock, and the appeal of Raytheon's idea for a special new "enhanced" Paveway—a weapon that might be even more effective at striking a hardened terrorist lair, or a cave complex somewhere across the world—became even more obvious. As plant managers tried to accommodate an exponential increase in production, Raytheon execs approved engineers rededicating themselves to the race for an "enhanced" GPS-guided bunker buster Paveway III.

5.

Central Intelligence Agency HQ
Langley, Virginia

Luis Rueda dispatched the first American operatives to Iraq in 2002 and tried to survive hypertension. His guys would be exposed and alone.

He tried to think about all that could go wrong only when it was helpful but failed at that. He slept in his office but didn't really sleep; took his meals at work but they barely passed as edible. He survived on vending-machine sandwiches that looked like they'd been behind the glass since Dulles ran the building, less food than fossil record. And when those ran out, he sustained himself on candy and nuts squirreled away in a desk drawer, which was apparently what half of Langley was doing—more and more Agency staff working long hours to support ever-expanding counterterror operations—because soon so many people complained about the CIA risking a rat infestation that the cafeteria expanded access to twenty-four hours. The Langley medical bay was becoming a battlefield triage tent; Rueda found himself down there with a blood pressure cuff on so often, he started charting the numbers so he could make over-under bets with colleagues.

It began to seem to Rueda like the entire operations group was at stroke risk, maybe the entire Agency. It seemed like every other day ambulances carrying young people with old-people problems left the building, the CIA becoming a laboratory for errant vital signs and sedentary illnesses. He felt less like he was working at an elite spy agency's HQ, more like he was trying to run ops out of a nursing home. The shadow cost of the war on terror, Rueda figured as he walked the halls, might be this, right here: the epidemic of dorky wrist braces and compressed nerves at Langley.

He spent the mornings wading through cable traffic to distract himself so he might obsess a little less over his men who would just now be landing at a base in Turkey, rendezvousing with escorts the Turkish government insisted they allow to travel with them; just now arming themselves with assault rifles and sidearms, climbing into jeeps and Land Rovers loaded down with cash and explosives, rumbling over the Khābūr River and crossing into Kurdistan, passing between the sheer rock faces of the Zagros mountains.

A million things could go wrong. They could tumble off the road. Rueda had devised ways to hide explosives and cash he was halfway confident in—inside MREs, the "meals ready to eat" that the team carried, and behind the lining of heavy-duty Pelican cases protecting their communication equipment. They would pass a first glance but probably not a second. What would the cover story be if eight dead, fully strapped CIA operatives were found in a fiery pile of cash and weapons at the bottom of a ravine in a country Rueda's government didn't yet admit even to allies that they were operating in? And if they were seen, they wouldn't be able to even try diplomatic cover. If anyone anywhere would believe a bunch of burly bearded men with a small arsenal saying they did political affairs or something at the embassy, surely no one would in Iraq, where the United States didn't have one. Rueda flitted around Langley, working, mostly worrying, wondering why he hadn't heard from his guys, knowing exactly why he hadn't heard from the team.

Then, finally, a message from Tom: We're in.

They'd made it, they were in northern Iraq, everything smooth so far. They'd found a suitable place for a safe house, a mountain outpost at 6,000 feet; they were up and running—a little impatient with their escorts from Turkey, who, Tom reported, spent most of their time watching Turkish porn and smoking, but otherwise fine. Rueda's guys were ready to begin chumming the water.

As word filtered out to the Kurds around the mountains of northern Iraq that there was a group of men offering cash for intel on Saddam's regime, responses from interested parties began to trickle in. Tom began handing out Thuraya satellite phones—cool toys with the added benefit of being trackable—to promising leads, but most of the leads claimed knowledge they couldn't possibly have.

One came with a vial he said was poison from one of Saddam's weapon plants, then spilled it on himself. Even when the intelligence checked out, it was usually old. It soon became apparent Rueda's guys weren't getting anywhere. One claimed to be a mechanic in Saddam's air force. Rueda read Tom's report and cross-checked the mechanic's intel. Some checked out but most of it was old. Rueda read Tom's cables showing only more problems. Radical Islamists were beginning to gather in the same area his team was trying to operate in, having fled the American invasion in Afghanistan, which meant they faced even more threats than Rueda already thought. There was chatter about a possible defector threat. The intelligence was thin; thin enough he could make himself believe it probably wasn't true. But if Saddam somehow suspected Americans were operating in Iraq, sending in a dangle—an agent posing as a defector—was exactly the kind of thing he'd do.

So when Rueda got a cable from Tom in August 2002 saying that one of their new Kurdish sources had something valuable to share—"A religious group," the source had told Tom; "I think they'll be willing to help"—Rueda was skeptical. It was another headache. There was little reason to believe this particular Kurdish source was credible. He'd been complaining that Tom's team wasn't treating him well enough, and Rueda had a sensitive antenna for the kind of would-be sources who tended to inflate their own value once money was being thrown around.

The source claimed he had access to a sect of Sufi mystics who were uniquely positioned inside Saddam's security apparatus. The kind of moon-and-stars offer Rueda now reflexively questioned. And it wasn't just too good to be true; he also found it really fucking bizarre. Sufis were something like the hippies of Islam, but this particular subset—they called themselves Kasnazani—were even more unique. The Kasnazani claimed to be capable of such extraordinary ecstasy that they would literally injure themselves. They'd pierce their faces with pikes, chew blades, and swallow sharp objects; some of them electrocuted themselves. They saw themselves as living proof of the supernatural, so their devotion to the Divine—and to the leaders on earth who opened a portal to it—was absolute. And they followed a kind of first family of leaders, sheikhs descended from the movement's founding prophet.

The kicker was that Saddam had apparently once trusted them. For decades, so the story went, this first family had enjoyed a kind of symbiotic relationship with Saddam's regime. They helped him maintain stability in the otherwise restive northern region in return for business opportunities and favorable jobs in his government. The relationship lasted for decades until a falling-out sent the first family fleeing back into the mountains, with much of the Kasnazani sect following. But not all of them. Some Kasnazani, especially ones with decent jobs in the government, stayed in Baghdad. The Kasnazani were more a religion than an ethnicity—it wasn't like they had easily identifiable physical features—so many just stopped publicly practicing the rituals that gave them away, kept their heads down, and climbed the professional ranks. Some of them, the source claimed, now worked in various arms of Saddam's security apparatus even as they maintained feverish—but invisible—loyalty to their sheikh.

Rueda tried to suppress his skepticism. The story the source was telling effectively described a ready-made sleeper cell. Could there really be a preestablished network of spies penetrating Saddam's security apparatus, loyal to just one family? It was like an entire building wired to a single light switch, and all Rueda had to do was figure out how to flip it. It couldn't be that easy. Rueda's experience in covert work was that when something seemed too good to be true, it usually gets you fucked over, and if the disgruntled source spinning this tale was even halfway savvy, he'd probably know how to attract a spymaster's interest. This was, in fact, exactly the kind of story a "dangle" would tell.

But Rueda's men were getting nowhere. If the Bush people decided they wanted to start the invasion, he'd have no useful intel to give. He didn't feel he could reject even a long-shot chance.

He cabled back to Tom: Your call.

Tom said he was going to pursue it. Which, for Rueda, meant even more waiting and worrying. More time trying not to visualize the worst. He went to more White House meetings. The Bush people made his head spin. Senior administration officials began to give what looked to Rueda like an aggressive display of misunderstanding the country they'd been desperate to invade for years—and an almost comic disregard for the advice of anyone who actually knew what the fuck they were doing. He watched Tommy Franks, a decorated general with a chestful of medals, trying to convince the Bush people that they were going to need something like 250,000 troops, while Donald Rumsfeld was saying 70,000 should be plenty, which was insane, then offered 140,000, which sounded to Rueda like a bullshit number. The serious people were trying to plan for a war; the political people were haggling as if the stakes were no higher than a good deal at a used-car lot.

Maybe troops felt like a step backward in the age of precision warfare, but when Rueda had brain cells left over to worry about anything other than his teams in Iraq, he worried that the people around Bush were missing the point. If the United States played its cards right—if they went in with overwhelming force—he could see large swaths of the Iraqi military dropping their weapons and defecting without a fight. But if it looked like Saddam had even a remote chance of holding on to power, he'd still be the greater threat to most Iraqis. Rueda didn't see people defecting en masse with half the necessary U.S. troops on the ground. If Rumsfeld was undercutting the generals on something as fundamental as troop strength, it was hard to imagine it all panning out well, even if Rueda was able to feed them gold-plated intelligence, which he was beginning to think might be as much of a fantasy as anything the Bush people were telling themselves.

Until one day, retreating back to Langley after more maddening meetings across the river, Rueda got a call from Iraq on the secure line.

"Holy shit, these guys are plugged in everywhere."

The Kasnazani tip looked like it might actually be legitimate. It didn't look like a dangle; it wasn't some clever counterespionage ploy by Saddam. Tom said he'd met with two brothers from the Kasnazani first family. The brothers wanted money and a guarantee that the United States was really going to invade, plus a seat at the table after the invasion, but they seemed to be for real. Tom said he'd played a little hardball—he'd told them that, before making any promises, he'd need proof they could deliver the kind of intel they claimed they could—and a few days later they returned with a brigadier general from one of Saddam's air bases in the trunk of their car. Under pressure from the brothers, the general began to detail technical specifications of aircraft under his command. Tom was impressed, agreed to pay the brothers a $135,000-a-month stipend, and asked how many more there were like the brigadier general. And, shit, did they deliver. Tom said he was sending Rueda a list with every member of the Kasnazani sect currently serving anywhere in Saddam's security apparatus.

Rueda jiggled his mouse, woke up his office computer, pulled up Tom's cable, and his eyes bugged out. The Kasnazani were everywhere. There were members of the sect all throughout the military, in Saddam's Special Security Forces, in his "Men of Sacrifice," the Fedayeen Saddam, in the country's intelligence service. If the list was legitimate, it meant Saddam's security apparatus was totally penetrated by this sect of self-mutilating mystics.

Rueda had stumbled on a case officer's dream. He began working to corroborate the intelligence from Kasnazani followers while Tom kept meeting with the brothers, who kept showing up with more well-placed officers in the trunk, and eventually with their father, the sheikh himself, and soon the Kasnazani first family was smuggling three or four people a week up to the mountain safe house. Tom was doling out a million dollars a month, disbursing Thuraya satellite phones like Halloween candy to dozens of promising assets the sect presented to him, and marveling at the influence this one family had. It was the craziest thing. When the smuggled-in sect members saw the sheikh sitting there, they lost it. Some started shaking. Some of them prostrated themselves before the sheikh and actually kissed his feet. And whenever any of them were hesitant or resisted giving information they felt was too dangerous to share, the sheikh simply cut them off, said, "You will cooperate," and they did. The team took to calling him "the Pope." Rueda had never been part of something like this. It was such an unprecedented, extraordinary intelligence bonanza—actionable, specific targeting intel Rueda could feed the military—that they gave agent reports the cryptonym DB/ROCKSTAR, and the volume and value of the intelligence coming from them proved almost sufficient to distract Rueda from an internal clock he could hear ticking ever more loudly.

He'd always known he had limited time before his men were found out. That timeline was now shorter because now there were Iraqis moving under Saddam's nose too.

Iraqis who were sticking their necks out because Rueda had promised them an American invasion was coming. And Rueda still didn't know when that would happen. His meetings with the administration weren't offering much clarity on actual plans. There was plenty for him to worry about without worrying about things he had little control over, but he couldn't ignore the growing notion that the Bush people—who'd never struck him as particularly thoughtful on the matter of Iraq and who hadn't yet given him any indication they were really ready for war—might not actually be as committed as they'd claimed to be.

He tried to ignore the sense threatening every day to send him down to the medical bay for another date with the blood pressure cuff—the idea that he might be right now helping history repeat itself. Now that assets were in play, he knew what it would look like if the invasion didn't happen soon. He could see it. He *had* seen it. When he closed his eyes, he was reliving his own origin story. Only this time he was Kennedy. Gathering up a brigade of exiles, sticking them on a boat, and shoving them off into the Bay of Pigs to do America's dirty work. Talking a bunch of suckers into believing they had American loyalty, which by now, Rueda knew, was usually an illusion. He closed his eyes and could see what it would be like for this strange Sufi sect. He could see what would happen to the Kasnazani if they were found out. Razed villages, sect members flayed and tortured, families raped. People suffering the most horrible fates, for the sin of doing what America asked them to do. Making the fatal mistake of believing him when he promised an American invasion was coming to back them up. Which was a promise, he now saw, that he didn't actually have the power to keep.

With assets now in play, he went down to the White House to find polite ways of telling them that now lives were on the line and they'd better start the invasion soon.

6.

Al-Fanar Hotel
Baghdad, Iraq

Kathy Kelly was trying to stop an invasion.

She stood out on the balcony of a small family-owned hotel in Baghdad her team was camped out in, trying to find news on a shortwave radio. She had some hope that cooler heads would prevail. The hope was beginning to thin; she knew people in Washington's halls of power were assembling their plans and excuses.

She fiddled with the radio and a voice crackled into focus. It took a moment to place it as Colin Powell. At a United Nations address, his slow voice just audible above the generator hum around her. Powell there for America, laying out each piece of evidence. She listened, almost sold. He was convincing even to her. Maybe Iraq did have weapons of mass destruction.

So this war might actually happen. Around her she already saw future casualties. She'd watched from Chicago as the promise of a warless new millennium slipped into memory, and once talk of another war in Iraq had reached a fever pitch, she joined another Peace Team. By then she knew a war would likely come as much from the air as the ground, and she knew who to call to get visas and fix the logistics, how to walk an unbeaten path to a place few were traveling to.

And now she was on the ground in Baghdad, hoping her own country would come to its senses, but putting herself here under the bombs just in case.

Her tactic now was finding real people, with real flesh and blood and voices, and sending their stories back to America in the hopes that faces and names might somehow make their way from the friends taking her calls in Chicago to some decision-maker's office in Washington.

She went around Baghdad escorted by the same small man with good English, a former spy for Saddam's intelligence service. Taking her requests and finding exactly the thing she thought she ought to see.

He took her to fasts in front of the UN headquarters in Baghdad and on visits to hospitals, visits to emergency rooms, tours of museums, so she could send back stories of friendly people, of wounded people, of clever, foolish, selfish, normal people, who might be harder to bomb.

She toured a music school in the city. What better place to find stories that might move Americans? She walked through a small hall with children's artwork on the walls, and halfway down she flinched at a picture. A child's scrawled drawing of a jumbo jet crashing into the World Trade Center towers.

She made her voice sweet and asked to meet the artist. She was soon surrounded by a gaggle of children.

"Allah wanted this to happen to people in *amrika*," an eleven-year-old said, "so people in America understand what happens when America bombs other people." A teacher showed up. The child smiled and shifted.

"We love the people of *amrika* and we want to be their friends!"

Kathy smiled back at the boy. "I was in New York that day," she said. The children did not quite get quiet. "And after that day, we met a lot of people who said they didn't want to see war. In fact, some of them, when they buried their relatives who'd been killed on 9/11, sang a song about how people are all a part of one another."

As Kathy's escort translated, the children looked at each other. "*Madam*," they said, and then they yelled, "WHY YOU DON'T TEACH US THIS SONG!"

The teacher, a kind, tired-faced man named Hisham, let out a laugh. An American in Baghdad teaching peace before a war? But he was both the school director and the city symphony conductor, so he couldn't sit on the sidelines. "OK," he said to Kathy. "Sing it for me."

And she did, in front of an audience of giggling children, while the director reverse-engineered an arrangement, and soon Kathy and the driver, even the former spy, were all sitting around a piano, singing their peace song.

She left, smiling, and the next day, received a summons to come back, and when she arrived the children sang the whole song back to her in Arabic, then sang it again in English, "*This is my home, the country where my heart is . . . but other hearts in other lands are beating*"—along with a new violin part the school director–symphony conductor had figured out, and a flute trill, too, recording it all on a cassette. And when it was done, the children smiled for approval, and Kathy applauded. This might move people. Maybe there was still a chance to stop the war.

She heard about the protests back in America and in London. There'd been protests against Vietnam, she'd been part of them, but not like this, *before* a war even began.

She continued her work. She prepared. She taped windows so that if the invasion came, they wouldn't shatter. She helped make banners. She hung huge pictures of Iraqi children from the hotel balcony. She turned into a party planner, hanging posters around the city. On top of the electrical facility: TO BOMB THIS SITE WOULD BE A WAR CRIME; across from the UN building: NO BLANK CHECKS FOR WAR.

Here in Baghdad, she could see people beginning to unravel. She could see some responding the opposite way. There were people who wanted to see the war she wanted to stop: the hotel cashier, leaning forward to tell her he hoped it would happen; a teacher wanting to teach freely, released from the burden of dictatorship. These people told her quietly, carefully; you didn't know which rooms were wired and who might be informing for the regime, so perhaps there was more she didn't hear.

The family that ran her hotel expected the worst. The night manager packed as much food as he could into the hotel refrigerators, enough to keep the kitchen going for weeks if need be. Neighbors showed up, asking to shelter there. The halls filled up; parents tried to distract the children. Teens excavated board games from hotel storage and played all day, mostly Risk; the younger kids ran around the halls, giddy with fear and excitement. Kathy saw the alarm, so well tamped down by so many in the city, beginning to break through.

She bonded with Miladhah, a sweet young girl who absorbed all the fear and turned it into a game. Miladhah found a flashlight, imagined it was a gun, and shot Kathy, shot her own mother, shot herself, collapsed, then sprang up and hung off Kathy's neck. Miladhah's mother was horrified, but Kathy found herself encouraging Miladhah's make-believe. Maybe it wasn't a bad way for the girl to cope with what she would surely soon see. To picture the worst, then imagine surviving.

But, privately, Kathy was unraveling too. Soon people would start dying here. She tried to occupy herself with preparations she knew were probably trivial: futzing with generators, checking survival kits, counting snacks, investigating the case of the missing M&Ms. Watching Miladhah dying and rising, dying and rising, giggling and crashing into Kathy's arms.

7.

Situation Room
The White House

Best case, civilians were going to die, General Tommy Franks explained, if the massive bombing campaign went forward.

Franks addressed the president and his closest advisors. It was now March 5, two weeks before the planned start of the war, and Franks was presenting a list of twenty-four targets he thought had to be hit because they were key to Saddam's command-and-control apparatus, but that were also "high-collateral-damage" targets. In populated areas, near mosques and schools and hotels. Even availing themselves of the uncanny intelligence from Rueda's Kasnazani network, even availing themselves of the best guided bombs in the arsenal—JDAMs from Boeing and the Paveways now coming off assembly lines at Lockheed in record numbers—and even if all those precision bombs landed at their exact intended aimpoints, a lot of civilians were still going to be killed.

As Franks spoke, Bush was staring down at photos of the targets Franks had printed out. "I see there's a school there," the president said.

"That's why we're going to hit the target at night."

Bush was quiet. He seemed to be thinking.

ACROSS THE RIVER at Langley, Rueda was more worried about the delay. The Kasnazani had been in play for too long already. He was the one responsible for convincing them an invasion was coming, and now he could feel Washington hesitating.

He went down to meet the National Security Council staff in the White House Situation Room with a mission to make them see the sect of mystics less as assets, more as human beings.

"Guys, they're trying to identify us, trying to identify our sources," he said. "There's no going back for us. If you stop, it's too late for these guys." He wasn't supposed to have an opinion on policy but, fuck it, he had an opinion. "If we don't follow through, people are going to die." He couldn't see if it was penetrating. "Saddam has better counterintelligence ability than we do," he said. "He'll start killing people to get to where he wants to get."

He waited. He went back to Langley to try and hide himself in cable traffic. He came up with new ways for Tom to say *Soon, I promise* to the Kasnazani.

Finally, on March 17, at 8:01 local time, the major networks preempted their regularly scheduled programming and President Bush walked out to a podium in the White House. He looked grave. Almost, at first, uncomfortable. "My fellow citizens. Events in Iraq have now reached the final days of decision," he said. "Saddam Hussein and his sons must leave Iraq within forty-eight hours. Their refusal to do so will result in military conflict, commenced at a time of our choosing."

Rueda watched from Langley. A "time of our choosing" meant 18:00 Zulu on March 21. "Alert Hour," 7:00 p.m. Baghdad time. It meant the massive bombardment was now set to go, planes with guided bombs waiting for the order, pre-positioned at bases all around the region, naval destroyers ready to spin up cruise missiles, many of the weapons destined for targets identified by Rueda's own team and the network of Sufi mystics.

But now that it looked like he'd have what he'd been pushing for, what he'd worried wouldn't happen in time, he didn't feel relief. He was out of emotions, bone-dry. He was either too tired to feel anything, or maybe whatever relief he expected to feel was outweighed by the fact that it was going to come from launching a war he wasn't sure the United States was actually prepared for.

As he waited, a cable addressed to all CIA stations and bases came through:

> In the very near future and absent some unlikely and extraordinary turn of events, our nation will embark on a dangerous mission to disarm iraq and remove saddam hussein from power

If Saddam was still in Iraq at dawn on March 20, then the bombing would commence soon after.

So Rueda at least had that. However shittily the war might go, however pathetic he believed the planning had been, and however insufficient he knew the troop numbers were, at least this time, it looked like we'd keep our promise to the people who were risking their lives for us.

02d:04h:30m to Alert Hour

President Bush made his way down to the Situation Room, where the National Security Council had assembled and was waiting for him.

Bush took his seat at the head of the table. A wall-mounted screen showed a beamed-in image of General Tommy Franks, now working out of an air base in Saudi Arabia.

"Can you hear me, Mr. President?"

"Yes, we can, Tommy," Bush said, "we can hear you fine. You've got the National Security Council here."

"We'll go around the horn," General Franks said. "I'll start with Lieutenant General Buzz Moseley."

Moseley spoke up from CENTCOM, where he was in charge of Air Operations. "Our command and control is full up, Mr. President." Moseley spoke in a soft Texas drawl. "Our Coalition Forces are in place."

Tommy Franks gave the floor to each component commander in turn, and to each the president asked, "Do you have everything you need?"

The Army ground commander said, "We're moving forward into attack positions. Our logistics are in place. We have everything we need to win."

The Navy admiral: "Green across the board."

The president asked, "Can you win?"

One by one, each commander gave a clipped confirmation. Then Franks came back on the screen.

"This force," Franks said, "is ready."

Bush turned to Donald Rumsfeld. "For the peace of the world," he said, "and the benefit and freedom of the Iraqi people, I hereby give the order to execute Operation Iraqi Freedom. May God bless the troops."

On that order, highly skilled teams of special operators stationed in the region began to filter into Iraq to set the stage. One team moved across the Jordan border to sabotage a missile site, another from the south to begin securing oil platforms, another from the north. Everyone moving into position.

Bush left the Situation Room and went back up to the Oval Office, but then went outside.

He walked slowly, head down, around the White House South Lawn. He prayed.

He came back in, sat down in the Treaty Room, and decided to write a letter to his father, who'd launched his own war against Saddam Hussein twelve years before.

"I know I have taken the right action and do pray few will lose life," Bush wrote. "Iraq will be free, the world will be safer. The emotion of the moment has passed and now I wait word on the covert action that is taking place."

He knew no matter how well it went, or how capable the military was, or how advanced its weapons, this was simply too big a campaign to avoid civilian casualties. But that was the price for toppling Saddam.

He didn't think it was an easy choice, but he didn't doubt it was the right one. "I know what you went through," Bush wrote. "Love, George."

What Bush didn't know—what no one knew: not General Franks, not Donald Rumsfeld, not any of the commanders—was that at that very moment, across the world, one of Luis Rueda's assets was initiating exactly the kind of unlikely and extraordinary turn of events that would give the United States a chance to call off the invasion of Iraq before it began and avoid everything that would follow.

8.

Special Security Organization Telecommunication Station
Baghdad, Iraq
Old: 24h:49m to Alert Hour

As President Bush was writing a letter to his father, just outside Baghdad an unassuming officer working at a telecom switching station noticed something strange.

Until now, he'd maintained a low profile. No one around him knew he belonged to the Kasnazani sect. No one knew he'd agreed—reluctantly—to be a secret asset for the Americans. The two brothers had asked him to come with them to a secret American outpost in the mountains up north; they'd clearly run out of more promising options for sect members to spy. When he arrived at the outpost, he tried to decline, but the brothers surprised him.

Their father—the sheikh himself—showed up, and the unassuming telecom officer couldn't figure a way to say no. At least there was reason to believe he wouldn't have to do anything. What useful intelligence could someone in his position possibly have?

But now, as he sat at his workstation, he was being confronted with a puzzle that seemed potentially relevant.

Up on the status board, a blinking green light representing the telecom network in one particular Baghdad neighborhood suddenly switched to blinking red: system failure.

Then, a little later, it switched back to green even though, as far as he knew, nothing had changed. No team had been dispatched to diagnose and fix the issue. The network had simply malfunctioned, then somehow instantaneously fixed itself.

A few days later, it happened again.

He started paying closer attention, since maybe there was a pattern here—but none emerged, and no one around him had any explanation. The network kept failing and kept spontaneously fixing itself.

A few days after he first noticed it, he saw the anomaly again, only this time he overhead a conversation a little later in the day that seeded an idea. Colleagues were talking about a neighborhood that someone said Saddam had visited that day, and it was the exact neighborhood where the switchboard status light had briefly indicated failure.

He started asking around every time he saw the glitch. He kept closer track of the news, looking for mention of Saddam's whereabouts. Sure enough, a pattern emerged: the glitch always corresponded exactly with Saddam's movements. The explanation, once the officer figured it out, was almost obvious: someone in Saddam's entourage must be tasked with disabling the network wherever Saddam went. Probably a security tactic to prevent anyone from reporting his location. But if the idea was to keep Saddam's whereabouts secret, then as long as the telecom officer was watching that status board, the tactic was actually doing the opposite.

He decided it was important enough to secretly update Tom, his American handler up in the mountains, who began working with other American intelligence agencies to corroborate the theory. The CIA sent coordinates corresponding to one of the switchboard anomalies to the National Imagery and Mapping Agency, which soon reported back: satellite imagery clearly showed a large convoy at those exact coordinates.

The theory was real, and that meant they had the holy grail—a read on Saddam's real-time location. If he could tell the Americans where Saddam was, some kind of precision strike was probably possible. That was the dream: decapitate the hostile government, and the war wouldn't even be necessary.

The once-reluctant telecom officer, realizing he was indeed onto something, was now even more motivated. He leaned in. He took it upon himself to recruit another Kasnazani he knew, who worked for a security unit that sometimes traveled with Saddam's personal detail; his CIA handlers gave the new sub-asset the code name "Rokan" and gave him one of the Thuraya satellite phones.

Almost immediately, Rokan passed on a tip. He said he'd overheard someone in Saddam's entourage say they were headed to Dora Farms, an estate along the banks of the Tigris, in the southern outskirts of Baghdad.

The unassuming telecom officer watched the big status board, and, sure enough, a light switched from green to blinking red on the part of the board that represented southeastern Baghdad. Someone in Saddam's entourage was shutting the network down exactly where Rokan said Saddam was headed. The telecom officer snuck out to call Tom.

9.

Central Intelligence Agency HQ
Langley, Virginia
Old: 21h:30m to Alert Hour

Rueda was snoring on a seventh-floor sofa when an urgent flash message appeared on his office monitor.

Worn-out and believing there was nothing left to do now that the massive air campaign was about to start, he changed into his baggy jeans and closed his eyes, so he missed Tom's message. Tom was apparently frantic: he'd decided everyone needed to see this, so he also sent the message to senior leadership at the CIA, almost the entire seventh floor as well as the CIA's liaison at CENTCOM, which was coordinating the massive air campaign set to commence in a matter of hours.

Rueda woke to CIA director George Tenet shouting at him.

"Come with us!" Rueda rubbed his eyes and asked what the hell was going on. Someone in the phalanx powering down the hall filled him in: They were going to brief Rumsfeld.

Rueda rubbed his eyes. *Fuck. I'm gonna need to get dressed for that.*

Tenet shouldered his way through the halls, trailing his deputy and a gaggle of staffers, Rueda trying to keep up as he wrested himself into a suit and tried to get up-to-date on the latest. There was already more:

Rokan seemed to be part of an advance security team, because he'd just called in to say he'd actually already arrived at Dora Farms. And he'd called the latest update in on one of the Thuraya satellite phones. Those were trackable. Sure enough, when Tom checked, Rokan's phone showed up on the tracking map near a bend in the Tigris River in southeastern Baghdad. Rokan was where he said he was. Rueda had two spies using two different methods to confirm Saddam's destination.

Almost immediately another flash message came in: staff at Dora Farms were apparently unpacking food and supplies as if expecting guests, and Rokan also reported seeing that the security presence at Dora Farms had increased.

As Rueda hustled to keep up with Tenet power walking through Langley, someone shoved a packet of satellite images into his hands. They were astoundingly clear. *I can read the fucking license plates.* One of the pictures showed Dora Farms with a time stamp from just a few minutes before that revealed a phalanx of white security vehicles—hidden but not hidden well—under palm trees at the complex. Something was happening *now.* Rueda jumped into an armored Suburban, his tie still untied, Tenet and the deputies piling in, too, the driver revving up and out of the Langley lot, pulling onto the parkway and gunning it toward the Pentagon. A chance to stop the war was within view but diminishing with each passing minute.

At the Pentagon, Rumsfeld needed only a few seconds. "We've got to take this downtown." Soon Rumsfeld was in an armored limo, Rueda and the CIA team back in the Suburban, the whole convoy racing over the Potomac toward the White House to try to stop the war.

10.

Oval Office
The White House
Old: 20h:26m to Alert Hour

Rueda crowded into the room behind a whole murderer's row from CIA and DoD. Even in the middle of this strange phase of his life, the moment seemed surreal, almost staged. He looked around.

Colin Powell looked ill.

Dick Cheney looked like he could eat someone's heart, but to Rueda he always looked like that.

Rumsfeld: *Is it just me*, Rueda thought, *or does Rumsfeld look giddy?*

There was Condi Rice on the Oval Office sofa, all refinement and poise, and the chairman of the Joint Chiefs of Staff, General Richard Myers, ribbon bar catching the light.

And here was Rueda, immigrant kid, refugee. *How the fuck did I get here?*

Before they could start, President Bush decided he wanted a different setting and ushered them out of the Oval Office and into the small dining room alongside it. "What's going on?"

Tenet got right to it, unfurling a map and the satellite printouts he'd brought along. "We think we have a chance to kill Saddam. We've got two guys close to Saddam. Saddam and the two sons have been here, and might come back, if they're still not there."

Rueda watched Bush closely—it looked like the president immediately understood the implications. Bush started to pace. "This is really good," he said. "This sounds good."

He paced some more. He asked how good the source was.

No one said anything. Rueda looked around the room. He caught Tenet's eyes. *He looks nervous*, Rueda thought. Then: *He looks like he's looking at me.*

Everyone was looking at him. Condi Rice had turned completely around to face Rueda. The president hooked a thumb in his belt and cocked his head. All of them staring at the kid with the missing bunny rabbit from the Bay of Pigs invasion.

Rueda understood, in some disassociated way, that this was an appropriate time to feel nervous. But he didn't. He didn't feel anything. He cleared his throat. He explained that the source was extraordinary. Intel from the Kasnazani network had been confirmed repeatedly over the past few months; as far as sources went, they were about as good as it got. "We'll never get a hundred percent confidence," Rueda said, "but the organization has proven reliable." He tried to think of how best to empower the president. "Right now, it's about 75 percent certain."

Bush nodded. He turned to General Myers, to ask the obvious follow-up question. If this was real—if they knew where Saddam was—how would they do it?

Myers said cruise missiles. They'd launch the million-dollar-a-piece self-propelled rockets from naval ships.

Rueda watched Bush think for a moment. The president seemed calm. He shouldn't be calm. This wasn't a target they'd been evaluating for months; they hadn't done a collateral damage assessment. It could be a trap. For all they knew, Saddam had packed women and children into Dora Farms, planted a tip, and was luring the Americans into starting the war off with another Amiriyah shelter bombing.

Plus, Saddam technically still had time to leave the country. What would it say about America if the president told the world Saddam had forty-eight hours and started bombing at forty-six?

But at that exact moment, at the safe house in northern Iraq, Tom was hustling to compose another flash message: According to Rokan, a yellow taxi had pulled up to Dora Farms and the passenger looked to be about six feet tall, with a thick moustache.

Also, Rokan had overheard conversations about a Hussein family meeting to be held on the compound around 3 a.m. local time. Plus, a Dora Farms staff member had been spotted holding a spool of what appeared to be some sort of communication line leading underground. When the asset followed, he found a hidden underground shelter, and a quick length-of-stride measurement allowed an estimate for how far the shelter was from the main house.

Tom compiled the latest and sent the message to CIA headquarters, where an operator saw it and, knowing its intended recipients had all just left, forwarded it to the White House Situation Room, where a staffer received it, took off down the corridor, and burst into the Oval Office to tell Rueda there was something he had to see *now*.

Rueda took the cable and read it, and even though he knew it didn't really work like this, not in real life—that this was Hollywood stuff, real-time intel coming into the Oval Office at the eleventh hour as the president made a decision—he cleared his throat again and drew the room's attention.

Rumsfeld, Cheney, Bush, Rice—the whole national security establishment fell quiet for the immigrant kid. Rueda shared what he'd just learned: it seemed his Kasnazani assets now actually had eyes on Saddam. And now they knew more than just the property. They knew the specific room he'd be in.

The mood in the room visibly lifted. The deputy national security advisor, Stephen Hadley, grabbed Rueda and the CIA deputy director and spread a map out on the coffee table. Two of the most powerful men in two of the world's most powerful security organizations dropped to their knees like children on a playdate and bent over the map.

"OK," Hadley said, "can you show me where the bunker is?"

Rueda dragged his fingers across the map, approximating the asset's foot strides, trying to convert fresh intelligence into target coordinates.

The three men pored over the map, decapitation within reach, the chance to prevent a war and save an untold number of lives, as excited chatter rose from all over the Oval Office, until General Myers stepped forward and said there was a problem. If Rueda's intel was right and Saddam was going to be underground in some kind of bunker, cruise missiles wouldn't work. They couldn't penetrate bunkers. Even ten or fifteen cruise missiles destroying everything on the surface wouldn't do much damage to a reinforced structure underground.

If indeed the intelligence Rueda's assets had provided was accurate and Saddam was underground, the general explained, there was really only one weapon that would work. But it was so new, it had not yet completed testing.

11.

Raytheon Missile Systems Assembly Plant
Tucson, Arizona

Like most of the country, engineers at Raytheon were waiting anxiously for the beginning of Operation Iraqi Freedom when they received word about a possible last-minute, high-risk, high-altitude mission against some kind of hardened target.

The mission called for bunker-busting capability, but because there wasn't time to suppress antiaircraft defenses first, it would have a higher chance of success if the pilots could bug out after weapon release without having to hang around in dangerous airspace, shining a laser down at the target until the bomb landed.

It was a target tailor-made for the brand-new "enhanced" Paveway III, the only bunker buster bomb that had GPS and INS in addition to laser guidance. But that presented a problem: the newest Paveway was not yet certified for use.

It had taken years of engineering—of failures, slowdowns, and then a post-9/11 acceleration—for Raytheon to finally figure out the first theoretically usable GPS-aided bunker buster Paveway III, and in the rush to prepare the U.S. military for the impending operation in Iraq, Raytheon had started delivering the new units to air bases around the Gulf just in case there was use for them at some future stage of the war; but the program was so down to the wire that they'd had to begin shipping the bombs before testing was completed.

Now a cargo plane spun up its engines and taxied toward the Raytheon factory. The final checkout assembly area was set right against Tucson International Airport. Once the test units were completed and loaded, the plane took off and flew northwest toward Edwards Air Force Base in California, where Raytheon had a team of retired pilots serving as "field support engineers" standing by to help Air Force staff at the contractor test facility.

The new weapon arrived at Edwards in a kit, so the test team put bomb bodies up on a stand, pried open the boxes from Tucson with all the tech inside, bolted the airfoil group to the back of each bomb, and the head with the new GPS-enabled brain to the front.

The test range ground crew wheeled the assembled weapons to the test plane and levered them up into its open bomb bay, connected the carriage hooks and lanyards from the plane into the bombs, plugged in the data cables, and now, for the first time, Paveway's brain was, in effect, connected to the plane's.

The Raytheon team briefed the test pilot on how to enter GPS coordinates into a portable data module, which the pilot then carried out to the test plane and up into the cockpit. He plugged it into a port, feeding GPS coordinates down to the bombs in the bay below.

The test pilot lowered the canopy and taxied while the Raytheon team and Air Force test officials gathered at the telemetry station.

Once at altitude, as the test pilot began the "consent" process for weapons release, a weapon delivery computer captured a last batch of data from the plane's infrared sensor and navigation system, updated the ballistic information, and fed an adjusted flight path down into the weapons' brains.

On the underside of the plane, the trapeze doors opened. Two Paveways dropped from the bomb bay, one and then the next, the lanyards unspooling and data cables detaching.

The lanyards maxed out and ripped from the bombs, triggering a squib inside each Paveway, which broke the seal on the gas bottle that powered the fins and woke up the thermal battery powering the seeker.

Both bombs began gliding through the desert air along smooth paths, making subtle adjustments toward waypoints, one behind the other, never bumping into each other or falling off course, never losing guidance, while the team down at the telemetry station squinted up. The bombs maintained their paths toward a spot above the target, guided by seekers able to switch automatically to GPS if the target became obscured, and to laser for "terminal homing" and maximum precision at the end of flight.

The two bombs struck at almost the same point, both burrowing through the surface.

Engineers at the test range were quickly able to certify a successful test. The new "enhanced" GPS-aided Paveway III had worked as intended, and dropping two at a time worked, too, which meant that so long as the pilots downrange tasked with whatever the mission was were willing to use a weapon that had never before been dropped in combat and had completed testing just a few hours before the mission, the new Paveways were cleared for use.

12.

Al Udeid Air Base, Qatar
Duty Station of U.S. Air Force 379th Air Expeditionary Wing
01d: 18h:17m to Alert Hour

At an air base about seven hundred miles southeast of Baghdad, stealth fighter pilots scrambled to get ready. The White House was considering a last-minute mission *that night* and Air Operations Command was assigning the mission to them. They weren't told what the target was, just that only one weapon—the newest, "enhanced" Paveway III, a bomb that arrived on base a few hours ago, before it was even cleared for use—could service it.

In a building near the flight line, pilots assembled before a computer console that held details of the mission. In a matter of minutes, intelligence had passed from Dora Farms in southern Baghdad across to the city center, up to northern Iraq, across the ocean to Langley, then across Washington to the White House Situation Room, up to the Oval Office—then *back* across the ocean to Air Operations Command in Saudi Arabia, down to the 8th Expeditionary there at the Al Udeid Air Base in Qatar, and now into a computer terminal. The intel from Rueda's Kasnazani network was converted into target coordinates in the computer system, then loaded onto portable data modules.

The two pilots said prayers and then each took his data module out onto the tarmac, up a ladder, and into the cockpit, where he plugged it into a port connected through an umbilical link down into the bomb bay and completed the last step in the Kasnazani intel's round-the-world journey, transmitting into the newest Paveway's brain.

It was now two hours from daylight. The pilots would be "comms out" from then on: antennas retracted, no signals coming in or out of the planes.

The pilots stayed in the cockpits, waiting for a signal from the control tower, which would use a light gun to flash the signal that they should begin taxiing—that a final "execute" decision had come down from Washington.

Oval Office
The White House
Old: 17h:27m to Alert Hour

In Washington, Luis Rueda was staring at the president. *I didn't vote for this guy, but you gotta give him credit.* Bush just looked calm. Too damned calm. *Maybe that's leadership.* He tried to picture himself in Bush's position. *My sphincter would be so tight, it'd snap a pencil.*

Rueda looked from Bush to the others. The whole room on a megadose of adrenaline. You could almost feel the altered pulse of it, a pulling current of invigorated blood moving through important people. Rueda was empty.

He tried to focus while the president mulled the decision.

Rueda ran through the new approach in his head. He understood the updated strategy: that now a strike on Dora Farms required a specific precision bomb and that the bomb could only be delivered from a plane. He pictured it. Pilots flying over Baghdad without protection. Human beings in danger. *Because of me.* The notion for a moment became vivid, electric, and he was up there with them, a couple of young Americans in subsonic planes.

Someone moved, the imagined pilots slipped from his mind, and he was present again, standing in an odd-shaped room in official Washington, feeling calm now and not entirely knowing why. Maybe exhaustion. Maybe because, save for that little blipped image of pilots, it didn't feel real. It was all too wild to be true. He thought of his father: "We're not immigrants; we're *refugees.*" Now here he was, in the Oval Office, standing with the cabinet, the people controlling the most powerful military in the history of the world, listening to Luis Rueda.

Or maybe the calm was because this wasn't war—not really. Rueda had *been* to war. He'd been in jungles as a junior officer; he'd fought and won the dirty wars in Latin America, he'd fought and lost the drug wars there too. He'd had his face right inside the violence, up close and personal with it. That metal scent of recently spilled blood, pointed gut pain from weeks subsisting on MREs; he knew the chronic body odor of jungle months gone without bathing.

This wasn't that. This was violence decided on and dispatched from an air-conditioned room by people in suits. He knew this was war but this didn't feel like war. This was war without the war.

The president drew everyone's attention; it was time to poll the room: *Was it worth risking these pilots' lives? Do we send the Paveways in? Do we try to stop this war before it starts?* He turned to Rueda. That soft drawl, composure and calm like he was asking whether Rueda wanted fries with his order.

"What do you think, Luis. Do we do it?"

Rueda plumbed for whatever reserve store of energy he could summon. He tried to focus. He didn't think Bush could actually be making him the deciding vote, but there was perhaps no more important time to give the right advice. He looked up at the president. "We should take the strike. It's a chance. If we're lucky and he's there when we strike, we may save a lot of lives."

Bush took it in. He took Cheney to the side to speak privately. He went behind the *Resolute* Desk.

Bush looked around the Oval Office one more time. And at 7:12 p.m. Eastern Standard Time—one day, seventeen hours, and eighteen minutes before the ultimatum was set to expire and Shock and Awe was due to commence—the president nodded.

"OK, let's go."

13.

Al-Fanar Hotel
Baghdad

The war started before Kathy Kelly thought it would.

In the middle of the night, she'd received a call from a journalist friend saying, *Forget what you heard: it's not tomorrow; it's today. It's right now.* He'd said he'd heard planes were taking off from a base in Qatar.

She gathered with the other hotel guests. They waited an hour, then two, and began to think the journalist friend must have gotten it wrong.

When the thunder finally began, it sounded tenuous. Like the explosions were muted, as if they were muffled underground. A rumble, a warning, then three more.

On that first night, she believed the people of Iraq would be fine.

The sun rose, the distant thunder passed, life went on. The loudest sound was now the chorus of morning birds chirping in the trees along the banks of the Tigris. Saddam was on TV, though someone said it was a body double, and it was a recorded message anyway.

On the second night the lights went out. Explosions thudded against a far part of the sky. In the hotel bomb shelter—not really a bomb shelter but a room with a few stacked prayer mats four steps down from the lobby—a minister on Kathy's team professed into the night. The teenagers moved their Risk pieces around, eyes yanked up by the distant drumming. The night manager watched his daughter Miladhah aiming a flashlight like a gun at her own chest, dying, collapsing into Kathy's arms, reanimating, springing up, laughing maniacally. They were frightened, but they were also OK.

After those opening salvos, though, it became a different war.

By the third night the explosions felt closer and more ferocious. Foundation-shaking thuds so that even walking the halls you could feel your insides vibrating. The night opened; blades of light stabbed through and stone exploded, closer now. The bombs rattled the walls, shaking up through bodies and into psyches, rearranging psychology. Kathy stumbled into the lobby and saw the accountant's daughter fixed to the floor, unmoving, standing bolt upright over a puddle. Looking across the room at her father, her father looking right back at her, rage vibrating in his eyes.

Miladhah became less prejudiced with her flashlight, waving it like a tommy gun now, murdering them all. The teenagers pretended to focus on their board games, looking up after each roll of the thunder. *But that one was bad, no?*

Kathy summoned her superpower. She sent herself somewhere else so she could lie with her face. She relaxed her jaw after each explosion and smiled, as if stillness, like fear, could be a spreading thing too. She tried to convince the kids they were fine, this was all fine, but she was working against the adults, too, and the adults kept giving it away. They clicked their tongues, they winced, they said, *La-ish, La-ish. Why?*

When the pace of explosions slowed, Kathy tried to play. She got the night manager's girls to giggle. She walked the halls with them, one under each arm like duffel bags, taking them on fake vacations. She tried to occupy their entire imaginations so they'd have none left for picturing the things happening in their city, but even after the bombers retired for the night, she started to notice a droning. A kind of wrenching that, once she heard it, she couldn't ignore; maybe something clunking in a boiler room. Three days after she heard it for the first time, she realized it was coming from nearby. She looked down and saw it was the girls, both of them, grinding their teeth. The bombing outside did something physically. It altered their bodies. The hotel accountant's daughter, so bright and precocious, kept losing control of her bladder. It might have been pride that sent her tiptoeing farther and farther from the grown-ups when the grown-ups were distracted, exploring alone or maybe hiding, finding hidden parts of the hotel. Finding back stairways, and then one night when no one was looking she found her way up onto the roof and then outside.

Out, unprotected, where the bombs were close enough, you could almost feel the wind come off them.

AFTER A WEEK Kathy understood the pattern. Always loud nights and quiet mornings. Once they picked up the rhythm, the hotel guests tried for normalcy. They held morning music sessions together—*The Phantom of the Opera* and "Hallelujah," the Leonard Cohen version—then quiet time together, then prayers. One of the children had a birthday and they decided a party was a good way to try for normalcy. They went out to the river, someone scored a bottle of Pepsi, a girl turned thirteen on the Tigris. They pretended to celebrate until a tremendous explosion somewhere too close drew Kathy's eyes to one of the mothers, whose face shifted from fear to something more unsettling, something like disgust.

Community began to break down. Hundreds of bombs fell. The looting began. First a few boys skittering through, then roving bands with wooden beams and bats. Kathy recognized one of the early ones, a gentle young man she'd met and liked before the bombing. Now she saw him going into stores, swinging at windows with a broomstick. Things were shaking loose.

The night manager and accountant cornered her. *There are no police*, they said, *there is no authority. Saddam's forces are nowhere to be seen and the American invasion doesn't seem to include any people. There is just bombing at night, looting in the morning. The hotel will soon be hit. If not by a bomb, then by the looters. People out there know there are foreigners in here, with foreign passports, and satellite phones, and cameras, other valuable foreign things.* Kathy heard their concern; she also thought about kidnapping. She knew there was at least one gun at the front desk. "If they come to take us hostage," Kathy said, "please, don't pull out the gun. We'll just go."

In the mornings they did their morning routines, song and prayer, before the smashing started again, a horde moving closer to the hotel. Kathy prayed that she'd be brave when she was taken.

Then one day in early April, a few weeks into the bombing, the accountant's daughter came running from the roof screaming words in English it took a moment to understand.

"Soldiers! Military! I saw from roof!"

Her father tilted his head—*The roof?*—but by then the rest of the guests were already running after the girl, following her secret route up to the roof of the hotel, where they stopped and watched, dumbstruck. Down below, it looked like dozens or maybe hundreds of vehicles, a giant beige snake, inching its way toward them. Bulldozers, Humvees, fighting vehicles, and armored personnel carriers, a line of them stretching as far as they could see.

As the convoy inched closer, the hotel guests—activists, Americans, Iraqis—all seemed to have the same idea at the same time. They got the tape player, brought it toward the ledge, and played Leonard Cohen with the volume all the way up. They unfurled their signs. The American ground forces arrived in central Baghdad to a soundtrack of "Hallelujah" and then "Anthem" and then, because there weren't many other options, *The Phantom of the Opera*.

The convoy rumbled to a stop down below, and men in aggressive gear eased out of their desert-colored trucks. They stretched, they considered the signs: COURAGE FOR PEACE . . . NOT WAR, and WAR = TERROR, and LIFE IS SACRED. The men squinted up at the pictures of Iraqi children printed on vinyl. The men looked at each other. They took out their cigarettes. They looked back up at the roof. *The Phantom of the Opera* blared.

One of the men in unform finally broke the ice. He shouted up: "Where you from?"

Another chuckled. "You guys Red Sox fans?"

Kathy heard a woman next to her say, "My God, they're just children."

"They look pretty thirsty, don't they?"

The woman dropped the banner. "Of course that's the right thing to do." And without another word, the two of them marched back inside the hotel, down to the lobby floor, collected bottles of water, and walked out the front door to greet the men. Soon a procession of Iraqis and peace activists were fire-lining bottled water and a box of dates Kathy had been keeping under her bed out to the armored personnel carriers, and one of Kathy's team members read from scripture as the men drank and ate.

AS THEY ATE and drank, they began to introduce themselves. They were Marines, 1st Division, first to make it into Baghdad. A unit with a reputation for courage and aggression, but to Kathy they seemed almost gentle. Some of the men entered the hotel, not as an invading force but as interested visitors. Some of them seemed fragile. These were not hardened killers from a certain kind of fervent reporting, the articles enthusiastic about American heroism; these were the young men, the kids—*Gerber babies*, she kept thinking—from the stories beginning to emerge about men and women underequipped in almost every way for what they were now being told to do. The stories of enlisted men ordering body armor with their own money, buying adult diapers because they'd heard if they saw combat some meaningful fraction would lose control of their bowels.

One of the Marines knelt down and put his hands in Kathy's.

They began to talk. They told Kathy they'd already broken things in this place, and someone would need to fix what they'd broken. They said they wanted to be a part of it.

A Marine named Harris told her in a quiet moment he'd never wanted to go to war. He had a daughter who needed surgery to correct a birth defect, and the Marines meant income and security and healthcare for his daughter. He pointed to his rifle and said he was relieved, "I never had to use this," but his eyes danced away when he said it, and Kathy knew Harris had more to say about that.

He came back to her later to explain something about a checkpoint, a car approaching too fast, someone in the back seat, a girl about his daughter's age, maybe. Harris put his head in his hands. "I should've said, '*Shoot the fucking tires.*'"

There were cheerful ones—Tom and Jerry from Indiana—"You can call us Hoosiers, ma'am"—showing her pictures of their families, but mostly the Marines seemed to want to unburden themselves. Another came up to Kathy to say he, too, had done something he couldn't quite square.

He'd busted into a house, seen fatigues on the ground, and started shooting, *We had to for our own protection, ma'am*, but then he said, *We didn't really know, did we? We shot everybody.* He drummed his hand on his forehead. *I hope it doesn't register here, ma'am. You know what I mean? I hope it doesn't register up here.* A corner of the hotel lobby became the chapel pews, and Kathy ministered to people who Had No Choice But to Risk the Ultimate Price, and soon, as the Iraqi hotel staff ran around preparing food for the Marines, she realized their new guests were becoming protective. The Marines encircling the fifty-year-old women who kept coaxing their stories out.

Even after the Marines moved on, tearing down a statue of Saddam out in the traffic circle as Kathy's team watched from the roof, and making camp in the banquet halls of a nearby hotel, they kept stopping by to check on her, and Kathy began asking them to help with things she'd heard about. A cholera outbreak in Hillah; a warehouse full of food and blankets that could use some protection. Word got out that Kathy's team had a kind of partnership forming with the Marines, and one day Hisham, the music school director, showed up to tell her the looters were circling closer to them. If the Marines could help, just send even a vehicle, one jeep, maybe just park it in front of the school? But when Kathy went to ask, the Marines said they were getting skittish about extracurriculars. *It's even more dangerous*, they said. *You really need to think about leaving now.* And they didn't know this Hisham. If something went wrong, they could be court-martialed. They couldn't help him, and she couldn't count on them to protect her much longer either.

She went back to Hisham empty-handed. He nodded quietly. He understood.

But he came back to see her again the next day. "They came." Hisham had a hard time getting the words out. "They burned all the sheet music. They broke all the instruments." The music school was gone now. He didn't have much else to say, but before saying goodbye he wanted to give her something. "This is all they left."

He handed her a tape player from the Norwegian news crew and told her to put the headphones on. She pressed PLAY and heard her own voice. Then the children, the song she taught them just before the bombing began. "*This is my home, the country where my heart is . . . but other hearts in other lands are beating.*"

She began to sing along with the children, but only for a moment. She stopped because Hisham was crying.

14.

Central Intelligence Agency HQ
Langley, Virginia

A month and a half into the war, Luis Rueda sat watching on an office TV as President Bush landed on an aircraft carrier for a spangly declaration of success, giving a victory speech under a MISSION ACCOMPLISHED banner. Rueda listened as closely as he could, though he already knew what he needed to know.

There was still some mystery, but by then he had a sense for what had happened on the ground the night they tried to strike Saddam and nearly stopped the war.

The first reports had held that the strike was a success. The new "enhanced" Paveway IIIs worked perfectly. They struck Dora Farms at their exact aimpoints, penetrating its foundation, and their delayed fuses triggered a half ton of explosives collectively, all underground. The pilots were able to leave the airspace safely, and the Paveways were followed by cruise missiles coming off Navy destroyers, striking the compound shortly after the Paveways detonated beneath it.

Then there'd been more good news. Soon after the explosions, one of Rueda's assets on the ground reported seeing a man matching Saddam's description being pulled from the rubble, dead or at least close to death, an oxygen mask over the trademark moustache and his face gone blue. The man was piled into an ambulance and raced away before the asset got close, but all indications were that they had their man. When

Saddam appeared on television shortly after the strike, intelligence analysts said it was likely a body double.

The news was all good, none of it was quite conclusive, they didn't quite have physical proof that they got Saddam, and with the entire military apparatus already set in an inexorable momentum toward war, no one was quite confident enough to stand up and stop it. So, at alert hour, with a sense that Saddam might be gone but no one quite sure, the war simply went forward. Three thousand precision weapons were deployed over a two-day opening salvo, and as the first days of the invasion stretched into weeks, it became clearer that they had, in fact, just barely missed killing Saddam.

Rueda soon learned that the Kasnazanis had been right: there had been a high-level meeting at Dora Farms on the evening of the strike, with high-level officials escorted by extra security, but Saddam hadn't been there when the bombs struck. In the last hours of Bush's ultimatum, it turned out Saddam was worried about money and had been meeting with his finance minister and bank officials at his presidential palace, which happened to be in the same part of the city Dora Farms was in. Saddam was either at Dora Farms just before the strike or, as his men scrambled to pack up and move millions of dollars' worth of currency, they passed near it, shutting down the telecom network there and inadvertently signaling the telecom officer.

The strike at Dora Farms ended up killing or wounding at least fifteen people, just missed Saddam, and just missed stopping a war Rueda knew his country wasn't ready for.

Still, for Rueda, there was at least some peace in the fact that he hadn't abandoned his assets. The Kasnazani had risked their lives for him, and the American military support he'd promised had actually come. He'd fulfilled his part of the covenant, and, if nothing else, there was still that.

But as more news emerged from the aftermath of the Dora Farms strike, he learned that the asset code-named "Rokan" sending intel from the complex hadn't managed to get away in time. So among the casualties was an asset killed when the United States acted on intelligence he himself had provided.

Rueda would have to live with it. With American bombs falling, at least the Kasnazani were less likely to be rounded up and tortured. That would have to be enough. Because already he could see signs that things were getting bad on the ground.

"In the battle of Iraq, the United States and our allies have prevailed," President Bush said on an office TV. "Operation Iraqi Freedom was carried out with a combination of precision and speed and boldness the enemy did not expect, and the world has not seen before." The bravado didn't bother Rueda all that much. Bush was just doing what each of his predecessors had done: claiming credit for another revolution in the effort to fight war more humanely. Bush went on, "In the images of falling statues, we have witnessed the arrival of a new era. For a hundred years of war, culminating in the nuclear age, military technology was designed and deployed to inflict casualties on an ever-growing scale. In defeating Nazi Germany and Imperial Japan, Allied forces destroyed entire cities, while enemy leaders who started the conflict were safe until the final days," he said. "Military power was used to end a regime by breaking a nation. Today, we have the greater power: to *free* a nation, by breaking a dangerous and aggressive regime. With new tactics and precision weapons, we can achieve military objectives, without directing violence against civilians."

Rueda turned it off. He'd seen the precision; he didn't need to be convinced of that. He'd also seen a reliance on it, perhaps overreliance, and, inside the White House, something almost like scorn for the notion of boots on the ground. He'd seen the troop numbers requested by generals be edited down by politicians, as if politicians knew better and held generals as retrograde, living in the past, failing to grasp the promise of modern war. Rueda wasn't supposed to speak up in those conversations; perhaps he should've still—it wasn't like decorum was all that important to him anyway. But now, it was a matter of time before he'd be escorted out of them.

Just after Bush completed his speech, a team of lawyers filed onto the seventh floor. Rueda had an idea of what they would say. He was only authorized to run covert operations in Iraq in support of combat operations, they explained, and Bush had just declared that "major combat operations in Iraq have ended." If there was, officially, no more war in Iraq, Rueda couldn't legally work on it anymore.

That was all right. He'd done OK for a refugee kid from Cuba. By now he'd lost his insider status anyway; he was already being iced out by Bush's inner sanctum. This was only an official ribbon on what he'd sensed already: once the bombs started falling, he could feel the Bush people, never very enthusiastic about dissent from the intelligence folks, now wanting to hear none at all. Rueda was on his way out. He'd be hearing less from his network now and could only hope that whatever it looked like from the news, it wasn't too much worse on the ground.

15.

Baghdad, Iraq

On the ground, Kathy Kelly watches a country atomize around her. A thousand Paveways fall. She sees precision bombs cause broad-spectrum chaos. Two thousand Paveways, 3,000, more than in Serbia.

In the newspapers, American commanders are quoted showing what looks like concern about what's starting to happen here. Bombs aren't finding enemies. The enemy seems to have just disappeared. That can't be true, of course: Saddam had military, and intelligence, and police loyal to him. They're all just—*gone.* Or maybe they're hiding, biding their time so at some point, in some form, they can come back. No one seems to know.

Kathy sees things beginning to spiral out of control. Pedestrians are attacked in broad daylight. One of her team members is cornered and beaten. Others are robbed. She sees that people are having a harder time finding food and water. Five thousand Paveways fall.

The looters turn into purer anarchists, but they can't throw punches at American warplanes, so they throw bricks at buildings. Everyone must be seeing what Kathy sees, an accelerating chaos. Trash piles up in the street. Sewage spits up from sewers. Paveways strike sanitation facilities and meaningful parts of the electrical grid. Seven thousand Paveways fall, and in Washington the generals worry more loudly about whether they need more troops, more actual people. Eight thousand

Paveways fall, it's unclear on what. The enemy has long since turned to smoke.

In October, six months into the war, the American secretary of defense sees it too late. He writes a memo to his staff. "Are we capturing, killing or deterring and dissuading more terrorists every day than the madrassas and the radical clerics are recruiting, training and deploying against us?" The memo leaks, the papers grab hold of the question, and it's streamlined by repetition. In the papers and on news shows, the Rumsfeld question becomes simply, "Are we killing more terrorists than we're creating?"

Kathy visits blood-slicked emergency rooms. The bombs have all gone crazy, wild and excitable, hitting everything now. They keep making mistakes, finding civilians. She meets an eighteen-year-old with his leg caged in an external fixator, bolts screwed through skin holding on to snapped bones. She meets a sixty-three-year-old man who was out shopping for his family when shrapnel coming off a bomb sliced through his intestine. She meets a thirteen-year-old with a colostomy bag. She visits a twelve-year-old boy while he's still asleep after a double amputation. He wakes up and nods at the bulbs where his arms used to be. His eyes fly across the strangers around his bed, looking for his father and mother, but they are not there. Just an aunt and this strange white lady with wild bright hair. The muscles in his neck move. He looks up to speak. "He asks," his aunt translates, "how long it will take for his arms to grow back?"

Nine thousand Paveways fall. So many people are dying that the invasion becomes an area of interest in academic medicine. This is some kind of plague. *The New England Journal of Medicine* estimates 151,000 people dead by 2006. An atomic bomb's worth, more than Hiroshima. So many are buried without official notice that a controversial *Lancet* article says it's actually more like 650,000. And a factional war is only just beginning.

Kathy and her team stay in Baghdad for as long as they can, a year and a half, until four activists from a related group are kidnapped and she knows that staying in Baghdad now means bringing more attention to herself than she's bringing to the human toll on the ground. Again she moves, only to Amman, Jordan, one country over, as if to remain on call, this time to the red-light district, where hotels are cheap. She stays there long enough that word gets around. In the unlit alleys, the lobbies, among the halogen cashiers sweating behind plexiglass, men tilt their heads as she moves past: an American, maybe rich, probably connected. She notices them noticing. She's not sure whether they're thinking kidnapping, revenge, robbery, something else.

One night she knows she's being followed. She quickens her pace, hustles without running to her hotel, crashes into the lobby, her heart thumping now, and hurries down the hallway to her room. She unlocks the door, swings it open, slams it shut. There are footsteps, getting louder, closer, coming down the hallway after her. There's a pause, then they're knocking. She tries to steady her breath. There's nowhere to go now. She can't stay in the room forever; anyone she calls to help is just as likely to be involved in whatever crime her pursuers have planned; and anyway, it's not a reinforced door: they can kick it down if they want. Best to face her fate now, try to talk them out of it. She turns, undoes the dead bolt, and opens the door.

Two men stand before her. Behind them the hallway light flickers. "I'm Fadi," one says. "His name is Jihad." They have something to tell her, they say, something important. We were in Baghdad, too, when the bombs started falling, but there's more.

WE WERE FOREIGNERS there too, Fadi says. We had traveled to Iraq for university. We lived in a high-rise dormitory, and we know we stayed longer than we should've, but we were so close to getting our degrees. And by the time the bombing started, and classes stopped, it was too late to leave. Fadi sighs, collects himself. He goes on.

One day the Marines stormed the dormitory building. A misunderstanding followed, and the Marines seemed to think they couldn't leave the two young men to their own devices. They were handcuffed, moved around; after two nights in Baghdad they were loaded onto a truck and driven four hours south, to an air base the Americans had taken over. That's where we were kept, Fadi said, for months. The air base had grown into a kind of tent city prison, and, just like in Baghdad, there were too few American troops. People even younger than us were inside, Fadi said, children, really, some were being beaten and even molested by other prisoners. There were attempted rapes. There weren't enough blankets to protect all of us from the cold, and there were no beds to separate us from insects in the dirt. We spent every night shivering, right on the ground, trying not to think about deathstalkers, scorpions the color of dirt, almost invisible even in the light. Trying to ignore the yelps of children, the sound of skin on skin. It took us two months to see someone in a position to let us go free, an American officer serving as a magistrate.

This place they held us in, Fadi says, is called Camp Bucca. And among the people disappearing into this prison camp were five friends of ours, also picked up by accident, still being held there. If an American goes down there, maybe you can get the American soldiers to let the five boys go?

Kathy makes some calls to others in the peace team and manages to get ahold of "capture tag" numbers for two of the young men Fadi and Jihad were worried about. Then she finds a car, forms a mini peace team detachment—two American friends and a translator—and they drive a few hundred miles across a country being rearranged by air strikes. They skirt the desert toward a town called Umm Qasr, all the way on the other end of the Arabian Peninsula. The desert gives way to streaks of black grease, and it's so brutally hot, they keep the windows down even though they're driving through swarms of bugs, zipping into window cracks and pelting their eyes, more swarms the farther south they go, so by the time the camp rises on the horizon, she's spitting insects from her mouth. It's January 2004, she's spent half a lifetime traveling to war zones, and this is the worst place she's ever been.

Somewhere a flag snaps. She doesn't know how they'll get in. A soldier appears and gets to talking. He tells them he's a dental hygienist from Tennessee, joined up after 9/11, but he can't wait to get this all behind him. Anyway: the camp is closed to outsiders, he says, it's just past visiting hours. Yes, he confesses, despite everything, there are visiting hours. He tells them to come back on Thursday.

Kathy lingers, considering options in her head, running the geometry of four prone bodies trying to sleep in a small car for five days. No Motel 6 out here, and can they make it back across the desert before nightfall? Her face must change as she thinks, because after a moment the dental hygienist takes a hand off his rifle and touches a thing on his uniform. There's a complex arrangement of cables on him. He cocks his head and speaks to his shoulder. A beat, a fuzzy shrunken voice crackling back.

Soon there's the rumble of a heavy engine as a boss approaches in an impressive vehicle, the de facto warden. Kathy knows what to expect from the person in charge of a place like this; she braces for some grizzled king wearing savagery as armor, excoriating her for the intrusion. But when the vehicle reaches them, the figure stepping down is a lovely young blonde in fatigues. She says her name is Major Garrity, she's in charge here, and she greets Kathy's small team with warmth. She hesitates when Kathy gives the names of the five young men and explains why the request is so difficult to grant, the scale of what's going on inside. "Just today," Major Garrity says, "we already processed five hundred new prisoners."

But the major wants to help, and soon she's figured out an exception, arranged for four of the young men to be found and brought to a visitors' tent, and produced a jeep to take Kathy's team the rest of the way over the sand into the prison tent city, to Compound 11, Tampa 11.

When Kathy walks in, the four young men immediately begin testifying to the abuses going on inside. Their account matches what Kathy was told back in the doorway in Jordan, down to the color of the scorpions.

Kathy keeps an eye on the guards as the prisoners detail the horrors inside, waiting for objections, but none come. The guards don't even seem defensive, except for a moment when one interrupts to say they're trying their best. They even agree to recommend the young men for release, but no one is under any illusions that the guards have any say, or that there's even a recognizable system to process people out.

So Kathy promises the four in front of her that she'll do what she can. She promises to raise their case to politicians in America, to the Red Cross; she'll call the embassies on their behalf. People need to hear about them, she says, and people will. She says it knowing she might not be able to do anything and that even if she does, getting them out of this place won't free them of it. That what's happening here might echo through these lives, and maybe beyond just these lives—that it might be trauma one body can't absorb or one generation even; there will be effects none of them can see yet.

When her hour is almost up, she asks them if there's anything else she can do for them. "Please," one of them says, "there are many here. Help us all."

She knows it's too much. That even just the need she sees, just in this one tent, might be too big to meet. She leaves, knowing even more acutely how much there is that needs fixing.

JUST BEYOND WHAT she can see, a few tents away, a shy man in his early thirties is just then being processed into the camp.

Olive-skinned and apprehensive behind wire-rimmed glasses, his name is Ibrahim, a doctoral student, not inclined to fight. Picked up almost by accident, in the wrong place at the wrong time. Ibrahim doesn't like violence at all. He is the kind of person—timid, studious—who's supposed to be eaten alive in prison; even a normal prison that was well run and adequately staffed. The facilities he's being ferried through in Iraq, the one up near Baghdad called Abu Ghraib, now this one in the middle of nowhere, are undermanned and overrun, the kind of unsupervised environment where people like Ibrahim tend to get picked apart.

But as Kathy leaves Camp Bucca, Ibrahim is developing the sense that he may have another way to maneuver.

He knows enough to talk to the different kinds of detainees. He recognizes in himself a gift for diplomacy. He finds he can reach across the different groups who've been thrown into the prison camp with no reason to get along and long, hardened histories between them. Two groups, in particular: the prisoners from Saddam's regime, and religious fundamentalists the regime was always at odds with.

When you look a little closer, he finds, they may have reason to hate each other but they also each have something the other can use. The prisoners who served in Saddam's regime have practical skills. They understand how a military operates. They know how to organize, how to fight.

The religious fundamentalists have a drive, a cause, even in here.

And for all the differences between them, he can see they now have at least one thing in common: a captor. Both groups are humiliated every day by the Americans. Ibrahim begins to think that if he can focus them on a common cause, he can draw power for himself. If he can bring them together, he can create something unique. A different kind of movement, with him at its top. A religious movement, but one with military discipline.

He finds it's easier to draw followers the more radical his opinions are, so his opinions grow more radical. His followers become more radical. A whole sharia court system emerges in the tent city, punishing Western behavior. Soon Ibrahim is running something like a government within the prison.

In the end, it isn't so hard. The prisoners all have different hatreds, but every humiliation hardens the one they share. No matter what else they may want or believe, they're all suffering in an understaffed American prison while the country outside is disassembled from the air.

Eleven months after Kathy's visit, Ibrahim is let out. Prison staff have little idea what he's been up to inside.

But outside, he advances the project he started in detention. He takes on a nom de guerre. Ibrahim Awwad Ibrahim Ali al-Badri becomes Abu Bakr al-Baghdadi. He has an ambitious mission and a chance to realize it. Together with other members of the Camp Bucca group who've been freed because the Americans didn't have the staff to keep tabs on what was going on inside, Baghdadi goes out into the desert to realize a vision of military discipline fused to religious zealotry, an organization capable of providing for its members, capable of defense, perhaps more than that, of offense—of going out and attacking its enemies abroad, the way America does.

Not just an organization, then; what he envisions is more like a *country*, an actual state that spans the whole Levant, in Iraq and beyond, in Syria too. An Islamic state in Iraq and Syria. Abu Bakr Al-Baghdadi emerges from the prison camp and grows the movement that comes to be known as ISIS.

BOOK VII

SALMAN

(OPERATION UNIFIED PROTECTOR)

1.

Tyler, Texas

In the mid-2000s, more than a decade after Weldon Word retired, his old coworkers began to see one last outlandish vision of his coming to fruition.

One of his protégés, a tech CEO named Steve Roemerman who'd come into the industry working under Weldon in the TI defense business, found himself applying lessons learned during Weldon's heyday to a tech sector that was becoming ever more essential to both the market and to people's lives.

Steve now ran a tech incubator where, almost without realizing it, he'd begun channeling Weldon's old tendency to look at seemingly unrelated trends and predict what would happen when they came together.

As Steve oversaw the launch of ever more cutting-edge tech companies in his post-TI career—a time when soon-to-be-dominant companies like Facebook and YouTube, PayPal and Twitter, were finding success by anticipating and exploiting technological advances less evident to others—Steve was beginning to see that of all the times he'd watched Weldon Word forecast a future no one else did, perhaps no single prediction was as impressive as the one that was coming true now. An idea Weldon had once laid out in an impromptu conversation the two had before Weldon retired.

At the time, Steve's role with TI was strategy manager, tasked with tracking the company's investments and making sure its different projects all made sense together. But he'd been running into a problem.

The '80s were giving way to the '90s—that, too, was a time of exuberant innovation, and TI was involved in all kinds of technological developments. The Department of Defense had computers processing masses of data and communicating with one another using a military network, helping the push for a commercialized version of "internetting" and one that people were beginning to call the "World Wide Web." TI was involved in that project, in developing microchips to turn different types of media into signals that could be sent from one computer to another, as well as communication platforms those signals could be sent on, and the company was especially immersed in developing technology that would allow it all to be done wirelessly from small devices people could carry with them rather than just stationary computer terminals, so that someday soon people everywhere might send and receive media from the palms of their hands.

The mandate from Steve's bosses was that TI should be avoiding distraction and focus on the work of realizing a dawning mobile internet market. But the company still had its hands in dozens of other projects. TI had highly classified contracts with the intelligence community and top secret defense projects Steve wasn't able to learn much about apart from the fact that they were occupying valuable talent, they all wanted more R&D funds for "enhancements" beyond what was required by the contracts, and, to Steve, they seemed to be all over the place. There were TI teams working on infrared detectors, line scanners, side-looking radars, new terrain-following radars, radars that produced high-resolution images of the ground like a camera would but that didn't need light.

The early obstacles to GPS had been solved, and all kinds of satellites were now in orbit, including spy satellites, but they had a key weakness and TI had a team working on that, too: spy satellites were still effectively just expensive cameras orbiting the earth at a few miles per second, and the best solution anyone had come up with to retrieve the film involved ejecting it in canisters that had to survive the heat of reentry and then trying to catch them in midair or hope they survived impact. Film from spy satellites could be snatched by adversaries and was often lost.

TI was one of the companies that, along with others, like the Eastman Kodak Company, had people trying to eliminate the need to retrieve film at all by using an entirely different theory of photography: instead of letting light expose film, they were converting light into electronic signals that could then be transmitted wirelessly. If there was money to be made off the technology in the consumer market, Kodak seemed to have the head start, patenting a consumer version of the product called the "digital camera."

And while the work—at least the work Steve was able to learn about—was impressive, it wasn't obvious what, if anything, all the different imaging technologies TI was working on had to do with each other.

Whenever he asked project managers why TI was in these businesses at all, they all gave him some version of the same answer. They all pointed him to Weldon. It was Weldon who'd lobbied to move projects to Lewisville, where he could keep an eye on them, even if they were not in his domain.

Steve started to get frustrated. Weldon must have asked a dozen different project managers to report to him, they weren't able to tell Steve why, and it was beginning to get in the way of Steve's ability to do his job. After weeks of trying and failing to see the logic, he decided the only way to get his arms around it was to go before the master himself. He went to Weldon's office to make a plea for answers. "You gotta help me understand what we're getting into here," Steve said. "I'm being asked to invest money in these things but I can't tell why we're choosing to be in these businesses, and every time I ask what the rationale is, somebody says, 'Weldon wants it this way.'"

Weldon thought for a moment. He got up, closed the office door, sat back down, and set those bright blue eyes of his on Steve.

"I'm going to lay this out for you," Weldon said. "At some point in the future, someone with some kind of digital imager is going to be able to take a picture."

Steve was with him so far.

"That picture is going to be run through software that will say it's a picture of a new runway." A little hard to believe, but Steve nodded.

"That's gonna go into a targeting database. Then, some days later, maybe years later, somebody decides they don't want that runway there anymore. They pull it up and there's the targeting info, and you dump it into a smart weapon, digitally, and that smart weapon and the air crew goes off and gets rid of that runway you don't want."

Now he'd lost Steve. Weldon was talking about an almost entirely autonomous strike. Engaging a target where, aside from loading the weapon, warfighters would be doing . . . almost nothing. Engaging a target in a way that required very little of any person, really, beyond just deciding the attack should happen.

Steve managed, "And who else believes that?"

"That's the wrong question," Weldon said. "The question is: What other future is more plausible?"

Steve said nothing.

"*This* is the plausible future." Weldon wasn't done. "The nation's defense, even given the limits of technology . . . *this* is what the future looks like. And we need to be investing to enable this future, because we'll prosper as a company, and we'll make the world a safer place."

Steve couldn't figure anything technical to ask about an idea that even by TI's standards sounded to him a little like magic. So he didn't say anything. Even coming from someone with Weldon's reputation for making predictions, there was little real reason to think this one would come true. There was little evidence anyone was even interested in the capability he was describing, let alone any reason to believe it was even remotely plausible. All the capabilities Weldon described were managed by different government fiefdoms without any real coordination. Weldon was predicting a technological convergence beyond anything others had envisioned.

But as it turned out, it was only a few years before world events conspired to produce both.

It was 1993, and Bill Clinton's CIA director, a hawkish lawyer named

James Woolsey, was trying to solve a problem stemming from another major drawback of spy satellites: their susceptibility to bad weather. There was a conflict in the Balkans, and it was a new kind of war, where distinguishing civilians from hostiles was proving especially difficult. Some of the Balkan territory was under near-constant overcast, so there were long periods of time when images from spy satellites were essentially useless. Overcast wasn't much of a problem for gathering intelligence on major powers, since a quick snapshot during even a momentary break in the clouds provided pretty much all the information you needed about large military installations. But to identify militia members or insurgent groups who were often physically indistinguishable from civilians at first glance, Woolsey needed more than a single image. He needed to see what people were doing. He needed to know where they were coming from. He needed a way to watch an area continuously, even when the weather was bad.

Director Woolsey called up Abe Karem, an Iraqi-born Israeli engineer with an invention Woolsey thought might be useful now. Karem had a company that made lightweight, remote-controlled planes that could stay aloft longer than piloted aircraft. The company had since gone bankrupt, but he said he still had a few prototypes lying around. Woolsey purchased five of them for the CIA. The Defense Department caught wind of it and bought its own versions, and soon both agencies shipped their prototypes to the conflict in the Balkans, along with repurposed NASCAR transportation trailers outfitted with equipment necessary to fly the unmanned planes. Since the prototype's new mission was to stalk the battlefield, looking for combatants, some of the people read into the program started calling the planes "Predator." At first, Predator couldn't roam far; it had to stay close enough to the trailers to receive guidance commands via radio. But it had an early satellite antenna for transmitting photo and video, which meant that even though Predator had to be controlled from nearby, the images it captured could be seen almost anywhere in the world, almost in real time.

When the first Predator flew its first mission in the Balkans, Director

Woolsey sat in his office at CIA headquarters, watching a live feed. He could see the tops of people's heads as they walked, at that very moment, on another continent. Woolsey was amazed. And because the engine power required to keep an aircraft without a pilot aloft was so much smaller—less a jet engine roar than a buzz, or a "drone," inaudible from the ground—the people being videoed didn't seem to know Predator was there.

Woolsey saw a structure that looked like a bridge. He typed a message into a computer.

What are we looking at?

Four thousand seven hundred miles away, an operator in one of the trailers in the Balkans typed a response.

that's mostar.

Woolsey could hardly believe it.

We're coming up from the south

Woolsey typed another question.

What's that in front of it?

there's a guy starting to get on the bridge

What's he got on his head?

Looks like a big hat. You want me to zoom?

It was uncanny. Woolsey was effectively issuing instantaneous commands to an intelligence-gathering system in an active theater, and since the plane could stay aloft longer than a manned aircraft, they could now try to solve the problem of targeting in the Balkans. With the ability to linger, Predator could help distinguish combatants from civilians. The Predator drone quickly proved itself to be an excellent information-gathering tool. But at almost the very moment it proved its utility, it showed its limitations.

Drone operators could identify targets, but to the Americans in the trailers operating Predators, the Balkan geography and architecture scrolling by on their monitors all looked the same. The drone operators had early success finding targets, but when they tried to explain the location to pilots in planes carrying bombs, it was hard to point out which specific house to hit. Pilots started bombing the wrong buildings and sometimes just gave up after orbiting fruitlessly over hostile territory, trying to determine which of a few dozen clumped-together houses the drone operator on the radio was trying to describe.

The solution was obvious: the drone operator wouldn't need to describe the target at all if the drone had its own laser designator. The drone operator who saw the militant go into a building could shine the laser on that building, and then pilots with Paveways would just have to fly toward the target's general area and drop their Paveways; the Paveways would find their own way to the target.

By the time the Balkan conflict ended, the first Predators carrying laser designators had been tested and the notion of a drone in warfare had evolved from novelty, to effective but passive observer, to supporting actor.

Soon the concept shifted again. In August 1998 a truck packed with explosives in a parking area behind the U.S. embassy in Kenya and one at a gate in front of the embassy in Tanzania detonated almost simultaneously, killing more than 220 people and injuring more than 4,000. When the attacks were traced back to the terrorist leader Osama bin Laden, said to be hiding in Afghanistan, the CIA decided to send a Predator equipped with a laser designator to go look for him. There was an airstrip bordering Afghanistan it could launch from, but the equipment and personnel required to maintain and operate a drone made for a sizable footprint; it would have been hard for the operation to go unnoticed. The CIA looked for a solution and found one by having some of the team work remotely. Predator was given an upgraded antenna that could receive guidance commands relayed by satellite, allowing it to be controlled from farther away. Takeoff and landing were still too difficult to do from too far away, just as it had been in Joe Kennedy's time, so a skeleton crew at the airstrip near Afghanistan was still needed to get the Predator airborne. But once it was, a team housed at Ramstein Air Base in Germany, under a giant satellite dish, could fly it.

In September of 2000, an operator sitting in Germany controlling the Predator drone as it flew over Afghanistan spotted a man on the live video feed who matched bin Laden's description exactly. The Predator was in a position to illuminate perhaps the most important target in the world. But at that moment there weren't any armed airplanes nearby. By the time fighters carrying bombs scrambled from the closest air base, or a cruise missile from a ship in the Gulf could be launched and travel overland to Afghanistan, bin Laden would surely move, and all they would have done was tip their hand. Bin Laden got away.

When footage from the drone over Afghanistan was shown at a meeting of Air Force brass, someone wondered aloud why they didn't just stick a missile on the drone. If Predator could capture video of a target and designate the target, why shouldn't it also be equipped to fire on the target? Had the drone had that ability, bin Laden would likely already be dead, and whatever heinous new attacks he was dreaming up would be prevented. No one came up with a sensible objection, so the Air Force assigned a special unit to work on modifying Predator with a laser-guided weapon. Their first experiments tried equipping Predator with the Hellfire missile, a weapon built off the success of Paveway, much more expensive and with shorter range, but also lighter. Within a few months they'd carried out successful tests, and by early 2001 an armed, unmanned drone that would allow someone sitting on a base in Germany to find, designate, and also fire on a target like Osama bin Laden was ready for service.

As the United States prepared to deploy its armed version of the

Predator drone, a new obstacle emerged: the German government balked. Germany was not at war, and there were legal restrictions to lethal action conducted from its own soil. The Air Force improvised one final work-around. The powerful satellite dish at Ramstein was connected to a fiber-optic cable that ran across the Atlantic Ocean and was linked to dozens of other American bases. Once engineers found a way to convert data from that dish so it could be transmitted via fiber-optic cable, a Predator drone over Afghanistan could be controlled from any base with access to that fiber-optic line. Drone operators could launch weapons in Afghanistan from control stations at American bases—stations where they already had access to military computer systems, networks maintained by intelligence agencies, and databases with information about potential targets.

With that last hurdle cleared, the United States could now launch air strikes in which no human being was at risk, and in doing so, approached the pinnacle of a mission Weldon had been on since working on a bomb to drop the Dragon's Jaw decades before. A way not only to entirely eliminate physical risk to the warfighter but to dramatically reduce the role people had to play at all.

Just hours before the first armed drone was scheduled to fly over Afghanistan, hunting for bin Laden, hijackers took control of four American airliners, and the West suffered the worst attack ever on its own soil. Though the armed drone hadn't been able to prevent the terror attack, it began to figure prominently in the conflict that followed. In the early days of the "global war on terror," the drone program was so small that some generals involved in the air campaign didn't even know about it. But over the early years of the military operation in Afghanistan and then the invasion of Iraq, as the CIA and the Defense Department began to acquire and deploy more Predators, drones grew from a mostly unknown novelty to a small but meaningful part of American strategy. Soon an updated version of the Predator drone was phased in, with a more powerful engine and the ability to carry a larger payload. Now, in addition to the lighter but shorter-range and more expensive

Hellfire, drones could carry Paveway as well. The drone program expanded, and soon the Air Force had enough armed drones in operation that they dedicated an entire fighter wing to unmanned aircraft. By the middle of 2007, the 432nd Air Expeditionary "Hunters," stationed at Creech Air Force Base in Nevada, had logged 250,000 flight hours over Afghanistan and Pakistan.

As the military and intelligence agencies purchased and deployed more and more armed drones, other NATO countries saw their promise and the technology began to spread. The United States began exporting drones, and by 2011 France and the UK had begun flying drones armed with Paveways, too—just in time for a series of world events that would give them new purpose.

2.

Iraq

By 2011, Abu Bakr al-Baghdadi was seeing his own improbable vision begin to seem achievable.

The bookish leader who'd found a calling in the Iraq prison camp Kathy Kelly had stumbled across back in 2004 was now leading a terror organization accumulating power, influence, money, and even land—the territory under the organization's control would grow to cover 100,000 square kilometers and 11 million people. Two of the movement's other influential figures had just been killed in a joint U.S.-Iraqi raid, securing Baghdadi's position as the unambiguous leader, and doors kept opening for him.

As American troops began a drawdown from Iraq, they left behind a security vacuum Baghdadi could exploit, and at the same time a wave of popular uprisings that began with the self-immolation of a fruit vendor in Tunisia was spreading to Oman, Yemen, Libya, and even Syria, next door to Iraq. As it did elsewhere, the "Arab Spring" uprising in Syria shifted quickly from a moment of hope to something darker and more violent. When the Syrian leader cracked down, the protests there turned to rebellion, and in the chaos that followed, Baghdadi dispatched ISIS fighters to take control of Syrian territory. When his fighters captured the symbolic Syrian city of Raqqa—which, twelve centuries before, had been the capital of a storied caliphate—Baghdadi declared that his organization was no longer to be called ISIS. It was no longer an Islamic state of just Iraq and Syria; it should now be regarded as a caliphate with global reach, one relevant to—and able to recruit from—people all over the world. The Islamic State of Iraq and Syria became just "the Islamic State."

In less than a decade, Baghdadi had gone from anonymous prisoner in an understaffed prison camp to taking the title of caliph of all Muslims, prince of the Believers, successor to the Prophet Muhammad himself.

Growing his ranks became easier. He recruited terrified people from the territory he took over; he recruited people inspired by his movement from other countries. There was a growing body of evidence that this shy young man from Camp Bucca was actually delivering on his promise: he was building a utopia in the image of the grandest caliphates, a claim few others could make—something few other terror leaders even thought possible.

With his territory expanding in Syria, Baghdadi launched a new offensive back in Iraq. He took territory faster than anyone expected, perhaps too fast for his own good: within three months of launching the offensive, he already controlled so much land in Iraq that leaders of NATO countries felt they had little choice but to act.

In August of 2014, Baghdadi's fighters began to die in sudden explosions.

It wasn't just bombs falling from fighter planes roaring by overhead. It was explosions that seemed to come from nowhere. Bombs coming from invisible, inaudible aircraft as well.

Baghdadi began to lose territory. First in Iraq, then in Syria too. Neither country was safe, and as the campaign against his fighters accelerated, Baghdadi knew he was going to need a new safe haven he could send them to.

The Arab Spring presented an opening for him there too. As protests across the Middle East and North Africa turned into rebellions and tipped toward civil wars, Baghdadi looked for a place where a regime losing control of its population might allow the kind of chaos that would make it easier for his fighters to slip in and operate out of.

3.

10 Downing Street
London, England

David Cameron was thinking about Libya.

He watched a hopeful uprising turn into a rebellion, and as Gaddafi began threatening to attack the rebel stronghold in a city called Benghazi, the new prime minister tried to reconcile two competing impulses.

Cameron fancied himself a beacon of restraint when it came to using force, so he was not naturally inclined to favor some kind of military support to defend the rebels. He'd spent his early years in Parliament trying to establish himself as a counterweight to what he saw as America's overreliance on its military. He'd made a point of it, even using a speech on the fifth anniversary of 9/11 to level public criticism of American military overreach. He called it "unrealistic and simplistic," and he said, "Liberty grows from the ground—it cannot be dropped from the air by an unmanned drone."

But he was worried about Gaddafi. He knew what the dictator was capable of. Back in the '80s, Gaddafi sponsored attacks all over the world, including on British embassies abroad and on protestors outside the Libyan embassy in England. Then, two years after England let the United States launch Operation El Dorado Canyon from one of its bases, Gaddafi's spy agency retaliated by smuggling a suitcase full of explosives onto a jumbo jet flying between England and United States. The explosives detonated over the town of Lockerbie, Scotland, killing all 259 people on board and eleven people on the ground. The Lockerbie Bombing was by far the worst terror attack ever on British soil, and in the years since, Gaddafi's security services had carried out a similar attack against a French airliner, sponsored terror groups in South Asia, made extravagant arms purchases from Russia, and began developing weapons of mass destruction.

And for all Prime Minister Cameron's protests about the new American brand of war, there was an exception. Back during the Balkan crisis, Cameron was a young television executive, so he was in a unique position when footage of the conflict was broadcast to Western audiences. He'd been haunted by the brutality the Serbian dictator Slobodan Milošević had demonstrated; Cameron was haunted especially by footage from the massacre at Srebrenica. And for all his condemnations of American attempts to spread peace through the barrel of a gun, it was hard to dispute that once precision air strikes were actually used in the Balkans, they'd been effective. So much so that, for many world leaders, a question lingered after Srebrenica: Would the massacre have happened at all if NATO had launched the air campaign sooner?

And in Libya, as Gaddafi set his sights on what he called a rebel stronghold, the situation was similar: there was a feeling in Cameron's government that if they didn't do something, David Cameron would have a Srebrenica on his conscience.

But this time, there wasn't an eager ally in America braying for a fight. When Cameron talked to President Obama, he found Obama's attitude to be starkly different from Bush's. After years of the quagmire in Iraq, the United States was in a season of drawing down its military commitments the world over. American troops had suffered too much, Obama said, and, beyond that, he believed America's role was too often taken for granted by its allies. The United States was not going to take the lead in Libya.

To the surprise of many, David Cameron decided to make the first move.

He assembled a small coalition with Lebanon and France, made an appeal to the United Nations for permission to intervene in Libya, and with someone else now officially taking the lead, the United States signaled its support. Soon the UN passed a resolution allowing NATO to commence "Operation Unified Protector," approving the use of "all necessary measures" against Gaddafi, with the exception of a foreign occupation. Just before Gaddafi's forces surged into the rebel stronghold, Paveways began falling on Libya for the first time since 1986.

Almost immediately, British military advisors identified a problem: though early reports held that guided bombs were driving Gaddafi's men from key positions, they weren't doing so permanently. Gaddafi's forces would flee and then simply come back. There simply weren't enough Libyan rebels to hold the gains made by NATO air strikes. The UK needed more boots on the ground, but there was little political will to send them, and anyway, the UN had prohibited a foreign occupation. Which meant either the campaign would fail or manpower would have to come from somewhere else.

4.

Manchester, England

Salman Abedi and Abdalraouf Abdallah were teenagers growing up in South Manchester, England, when the footage of protests across the Arab world began to dominate the news.

Both were boys, though Abdalraouf was already beginning to fill out, becoming a fulsome young man with an intimidating mane of shiny black curls. Abdalraouf, the proud-looking older kid who already had adult responsibilities. Kicked out of the house by religious parents after coming home drunk from a party one night, Abdalraouf already had his own place. He was impressive to younger neighborhood guys.

He was impressive to Salman, a wisp of a kid who hadn't grown into himself yet. At sixteen, his proportions were still all off; other kids called him "Dumbo" because his ears seemed too big for his face. He made half-hearted attempts at rebellion, as if boldness might distract from his awkwardness. He tried shoplifting like he'd seen Abdalraouf do, he smoked weed, and he even tried selling it like Abdalraouf did. He looked for diversions.

The two were a few years apart at Burnage, an underfunded academy for boys that was so decrepit administrators argued over whether it was safe to remain open at all, and only agreed after winning a government grant to rebuild it virtually from scratch. Salman and Abdalraouf sat in classrooms with fallen-in windows and buckets catching leaks, then looked elsewhere for recreation and exercise because construction crews were piling waste in the playground. Burnage was a struggling institution in a prosperous area: a magnet academy drawing immigrant kids from the grimmer corners of the city into a nice white neighborhood whose own children went elsewhere. Salman and Abdalraouf were bonded by the contrast they saw every day. And they were bonded by the similar experience they were having with their families.

The boys both had parents who'd been forced to flee Libya and who'd followed what was by then a well-trodden path from Libya to the UK. England featured prominently in Libyan history: it occupied the country during World War II, had partially administered it in the postwar years, and still retained oil concerns there. England was one of Gaddafi's great rivals, and was the only country that helped the United States execute the preemptive strike against Libya in Operation El Dorado Canyon. England was often top of mind in Libya, and to immigrants around the world it often looked like the platonic ideal of a safe, cosmopolitan country where newcomers from foreign lands had a chance to weave themselves into society. But while London usually had the greatest appeal, it was also one of the most expensive cities in the world, so many families looking to resettle in a cosmopolitan place with a reasonable standard of living looked to the slightly smaller, slightly less expensive city of Manchester.

By the time Arab Spring protests erupted in Libya, the Libyan community in Manchester had grown into the biggest in the world outside Libya itself, and you couldn't walk to the market without hearing some new theory about the protests growing back home.

Salman and Abdalraouf watched the footage from Libya, they watched their fathers watching it, everyone around them invigorated by scenes of a once untouchable oppressor openly defied on the nightly news. The boys absorbed the sense from their elders that this Arab Spring might finally be the moment Gaddafi lost his grip on power. And if it was, it was a thing worth traveling back for, to witness firsthand. To maybe even take part in.

When Libyan families in Manchester who'd fled Gaddafi saw on the news that the British were beginning air strikes against his forces, they knew their interests were aligned. And soon after the air strikes began, a rumor began to spread in Manchester that the British government didn't mind the idea of Libyans in England going back to be part of the movement.

Then a rumor began to circulate that the British Secret Service was working behind the scenes to expedite British Libyans' travel back to Libya. And since the prime minister's campaign of air strikes happened to coincide with the school summer holiday in England, there were few excuses not to go. With little else to do, Abdalraouf and Salman decided to join the wave of young men from Manchester going back to Libya for the chance to help accomplish what their fathers had not been able to do: to finally drive Gaddafi from power.

SALMAN WAS A few months from his seventeenth birthday when he arrived. Abdalraouf was nineteen. Besides the odd mix-up in gang activity in South Manchester, neither had any experience fighting. But once in Libya they were both sucked up almost immediately by a rebel movement desperate for manpower. They were assigned to different divisions. Abdalraouf was put to work accompanying women and children trying to escape regime shelling by fleeing across the border into Tunisia. Salman was assigned to a team raiding the homes of suspected Gaddafi loyalists, trying to transform overnight from a goofy-looking kid to a fighter with adult-sized responsibilities. He did his best to look like he belonged. He posed for photos with rocket launchers and heavy artillery. He tried tramadol, the synthetic adrenaline that muted pain and made you feel you could run through walls, a painkiller that doubled as a fear killer, helping to accelerate the transition from sheltered adolescent to passable fighter for many newly arriving young men.

They tried to learn how to fight as the tempo of air strikes from NATO planes overhead accelerated. But even as NATO planes launched more and more strikes, neither Abdalraouf nor Salman encountered anyone from NATO countries on the ground. NATO countries didn't seem to have sent any actual soldiers. The odd American or British advisor would show up, but they always seemed to stay back from the action, as if afraid to be seen there.

So Salman and Abdalraouf found it was up to the Libyans to do all the actual fighting, the taking and holding of territory, despite the fact that they had no training. And the fact that the fighting kept getting more intense.

With less and less capacity to fire back at the NATO warplanes degrading his air defenses, Gaddafi ordered his forces to turn the weapons they still had against the people on the ground. Schoolkids arriving from England saw human bodies dematerialized by munitions designed for armored vehicles. Abdalraouf saw a friend standing next to him struck by some kind of heavy weapon, muscle and bone turning to mist. A heady summer vacation turned into something darker and more viscous. But they were freeing their motherland; they were making their fathers proud.

And if the kids coming in from Manchester were poorly prepared, still partly tourists here, they were also part of a movement proving itself effective. So much so that Gaddafi's son went on state television to claim the entire rebellion was a conspiracy hatched in Manchester, England. Murals appeared on walls in Libyan cities celebrating the "Manchester Fighters."

There was pride to help them abide the turn toward grittier fighting and the poisonous things they were beginning to see, and the bonus of knowing they were proving their worth to their adoptive country too. They could see themselves as working in partnership with the British government. They could see the British government in each bomb they saw destroy an enemy position they were facing, or piece of regime artillery threatening them.

And they were beginning to see those bombs from closer and closer, as the British began deploying a new weapon optimized for targets near friendly forces and civilians. Salman and Abdalraouf found themselves very near the receiving end of Raytheon's new Paveway IV.

5.

The British military had wanted a new weapon ever since they'd participated in Operation Allied Force to drive the Serb forces out of Kosovo. Slobodan Milošević had been compelled to surrender by a barrage of precision bombs, and American airpower experts declared the dawn of an airpower-only era, but British commanders hadn't been quite as sanguine.

As effective as the bombs in that campaign had been, they'd still sometimes missed. There'd been too many civilian casualties, and despite Paveway's usually impressive accuracy, laser was still too often thwarted by things as trivial as dust and smoke and even bad weather.

British commanders wanted a bomb that had no such limitations: a bomb with guidance as precise as laser on a clear day, as immune to weather as GPS.

But they also wanted a bomb suited to the new dynamics of war. England was also contributing to air campaigns in Iraq and Afghanistan, against enemies who rarely wore uniforms and sometimes hid among civilian populations. Enemies were becoming harder to find, which made them harder to plan for: the odds that a British pilot who happened to spot a terrorist would also happen to be carrying the right weapon for the target conditions were decreasing, and the time it took for some other plane with more suitable weapons to arrive could be the difference between eliminating a terrorist and letting one get away.

War planners began talking about "dynamic targeting": the idea of empowering pilots to engage enemies emerging in real time; enemies that you couldn't necessarily plan for. You couldn't load every plane with every possible weapon, and heavy, fuel-hungry jets were already going to have a hard time staying above a battlefield long enough to see, identify, confirm, and engage a target.

This new kind of targeting, for this new kind of war, required a single weapon appropriate for a whole range of circumstances—a single weapon that worked in good weather or bad but that also could adapt to different types of targets. One bomb that could destroy an entire cave complex or kill a single terrorist while sparing civilians nearby. A bomb capable of a big shock wave, or almost none; as powerful as previous Paveways but better than ever at avoiding civilian casualties.

And when the British Royal Air Force began asking around about who could build a new combination bomb, they found Raytheon was already partway there. Raytheon had already developed "dual-mode" guidance that could use both laser and GPS for its latest flagship weapon, the big, "enhanced" bunker buster Paveway III that the U.S. Air Force used in an ill-fated decapitation attempt against Saddam Hussein.

For that Paveway, Raytheon had built on Weldon's work at TI to develop a way for the bomb to orient itself as it fell. Ground crews could dial settings into the enhanced Paveway III so that when it landed it would be positioned at the near-vertical angle required to penetrate a reinforced surface.

If a Paveway could already do that, there was no reason it shouldn't be able to orient itself to land at some other angle as well. A bomb capable of landing in different positions would also be capable of different blast effects. And if the process of setting that angle could be improved, streamlined, so that it didn't take a ground crew—so that a pilot could do it, even while flying—then you wouldn't need to know the type of target before the mission.

Once Raytheon engineers saw it was possible to give pilots control over the nature of the detonation, they looked for other ways to let a pilot customize the weapon's function.

They developed a way for pilots to control not just the angle the bomb landed at but the direction the bomb approached from. Pilots who spotted a terrorist leader near a crowd of civilians could dial in a strike from a direction that would drive shrapnel away from them—even when that required the bomb to glide past the target, turn, and then strike coming back.

And to make sure the weapon could destroy underground targets—which required the bomb to delay its detonation—but also strike normal targets, and targets in especially high–collateral damage areas, engineers built a "height of burst sensor" into the new Paveway. Now, a pilot attacking a hardened target could not only set the bomb to detonate after impact but could also set it to detonate at the moment of impact for normal targets or even to "airburst" above the target.

The new concept for Paveway would give pilots an unprecedented level of autonomy. Its versatility allowed targeting decisions to be made more quickly, which meant they could be made further down the chain of command. The decision to use lethal force didn't require generals and war planners in command centers haggling over logistics and debating which plane carrying which munitions to send. The decision of how to strike—and whether to strike—could now be put in the pilot's hands and delayed right up to the point the pilot hit the weapon release button.

To further minimize civilian casualties, Raytheon engineers designed a new arming protocol. Paveway IV would fall from the plane "unarmed"—effectively, an inert projectile until it was a few feet over the target, when its fuse would begin communicating with its guidance system to confirm the bomb was within preselected GPS coordinates. If it wasn't—if the enemy was disrupting it with some kind of jamming signal or if it had been blown off course—the explosive payload wouldn't arm and the bomb would hit the ground without detonating.

As testing proceeded, Paveway IV revealed itself to be a weapon the British military would indeed be able to use in an impressive range of conditions and for nearly any kind of target. It was a culmination for Paveway: adding to and improving on all the most useful capabilities of each of its predecessors, all perfected, streamlined, and incorporated into a single package. And with its suite of methods intended to minimize collateral damage, the newest Paveway gave war planners confidence that they could fight wars even more humanely, even in difficult environments full of civilians and friendly forces. Paveway IV was a weapon nearly immune to restrictive rules of engagement and one with a compelling case that it was the safest weapon yet: safest for friendly forces near targets, safest for civilians on the ground, and safest, perhaps most of all, for the consciences of those deciding to drop it.

6.

Tripoli, Libya

Air strikes were now happening all over Libya, and as the regime reeled from the onslaught, Salman and Abdalraouf were dispatched by their commanders toward the capital for a final push to remove Gaddafi's forces from power.

In the skies above them, a reunion of sorts was underway: every version of Paveway in service was being used by eight different countries to pound pro-Gaddafi forces and clear the way for rebels. British pilots patrolling with Paveway IVs attacked a range of targets emerging in real time, destroying main battle tanks and armored vehicles, pilots cleared to engage targets that were almost preposterously close to rebels. They destroyed an intelligence building while barely damaging a hospital next to it. They attacked a four-story building near the seaport in Misurata but only the two floors that regime forces were using to fire on the seaport. Two Paveways dialed to a penetration angle and a delayed fuse setting pierced the roof, passed through the fourth floor and into the third, then detonated before reaching the second, destroying the top two floors while sparing the bottom two.

Enhanced Paveway IIIs destroyed the Gaddafi regime's command-and-control centers and a military storage complex built into a mountainside, using their controllable flight angle to attack from the side, approaching at a nearly horizontal glide through the valley, and then crashing through the front gate before detonating.

Standard Paveway IIIs made quick work of regime radar and missile facilities, while Paveway IIs struck howitzers, vehicle depots, and rocket launchers.

One by one, explosions were eliminating every tool Gaddafi had to fight, and by August of 2011 it looked like there might be a chance to entirely remove him from power.

He'd lost his most sophisticated weapon systems; he was losing track of his own forces as NATO bombs took communication nodes offline, and as rebels on the ground captured and held more territory, a final decisive push to take the capital and officially end the regime became the obvious next step.

Salman and Abdalraouf arrived in Tripoli as NATO accelerated air strikes around the city, setting the table for the amateur fighters to make their advance. As they had been elsewhere in the country, NATO precision-guided weapons were devastatingly effective at weakening Gaddafi's forces in Tripoli, but the regime forces that remained were dug in under orders to exact as heavy a cost as possible. They'd settled into elevated sniper positions, laid tens of thousands of mines, and set up the last of the artillery that hadn't yet been destroyed.

As NATO planes made their last bombing runs, the rebels advanced on Tripoli from three directions. And if from above the outcome looked certain, on the ground, the fighting was brutal.

Abdalraouf made it to the northern outskirts of the city, but just before entering he passed through an uncovered area. Before he realized his mistake a sniper got off two rounds. Abdalraouf collapsed, disoriented and unsure how badly he'd been hurt, but someone dragged him to safety and the advance continued without him. Salman managed to avoid the worst of the fighting, and by the end of August, Gaddafi had fled and the rebels were celebrating in Green Square. The capital had fallen, and winning Tripoli made an end to the entire war almost certain. All that was left to do was root out the last pockets of Gaddafi loyalists. And, most importantly, to find Gaddafi himself.

No one knew where he was, but as rebel commanders and NATO generals tried to decide where to focus next, a regional television station aired an audio address from Gaddafi. He didn't reveal his location, but he declared the capital had not in fact fallen, because he'd moved it from Tripoli. Gaddafi declared the capital was now a city two hundred miles down the coast called Sirte.

The rebels decided that was as good a place as any to focus their attention.

NATO commanders did too. At an air base in Sicily, a ground crew loaded Paveways onto a Predator drone.

Once airborne, the Predator was taken over by an operator stationed in Nevada, who flew it across the Mediterranean toward the newly declared Libyan capital.

With a thumb over the weapon-release button and a military intelligence officer standing by to consult a targeting database and call in other NATO aircraft if the need arose, the drone operator started watching Sirte from above.

7.

District 2
Sirte, Libya

Colonel Muammar Muhammad Abu Minyar al-Gaddafi—great emancipator of the colonized, "King of Kings of Africa," and Brotherly Leader of the Libyan Great Socialist People's Jamahiriya—hid in an abandoned housing project.

He was cut off from his family except for one of his sons, Mutassim, and cut off from his regime except for a small posse of advisors and guards.

As far as he knew, his location was still secret. Sending the taped statement to the television station had been a gamble—drawing more attention to the city he was hiding in had been a gamble—but there was a chance it would work as a clever misdirection. Maybe naming Sirte the capital would make the rebels less likely to think that was where he actually was. Either way, he'd needed to show he was still the country's leader, you can't run a country with enemies controlling the capital, and if it came down to it, there was reason to believe the people of Sirte would stand by him. He'd put this city on the map; it was the city of his birth.

But now he was running out of food.

He had scant intel on how the war was going. The little he could glean from his son's satellite phone conversations with people elsewhere watching the news was more and more worrying. For a while, the rebels didn't seem to know he was in Sirte. But now he was hearing that they were amassing around the city anyway.

His son came to sit with him. They'd lost dozens of tanks, and air strikes were still pounding their staging posts and communication facilities. They couldn't be sure how much longer Sirte would stand, Mutassim said. So the only option now was to try to escape before it was too late. They'd have to figure the rest out later; for now, they had to find a way to sneak past the rebels amassing around the city and get out while they still could. And he had an idea how to do it.

Mutassim began gathering patients from a temporary field hospital in the run-down neighborhood they were hiding in, as well as doctors and nurses and any regime loyalists he could find. They'd form one large convoy with his father in a vehicle near the middle; they'd leave before dawn and maybe sneak past the sleeping rebels. If that didn't work, if the rebels saw them, the convoy would be big enough that hopefully no one would know his father was in a middle vehicle, and with any luck it would look like a medical convoy anyway. They'd caravan out to the western edge of the city, link up to the major north–south highway there, and follow it out of Sirte.

Mutassim worked through the night. He managed to assemble more than 250 vehicles, mostly regular 4x4s but a few mounted with antiaircraft cannons and machine guns just in case. He ordered the men to hide ammunition in any vehicle with space for it. The whole thing was an impressive operation for being so last-minute, but it took longer than he'd hoped: the sun began to rise, and by the time they were ready to leave, they'd entirely lost the cover of darkness. And Mutassim hadn't counted on the fact that there might already be eyes over Sirte. Or that, from the perspective of a drone, a few dozen vehicles leaving a civilian neighborhood at the same time would be a noticeable departure from the normal pattern of life.

A little after 8 a.m. the convoy began to move. For a while it snaked west along the coast without incident, but in full daylight it was impossible to avoid attention, and soon they were taking fire from rebels on rooftops. They made it out of District 2 mostly intact, then all the way to a last open stretch of land before the highway, and were almost out of the city when the convoy came undone. One of the lead drivers near Gaddafi's car slammed on the brakes, and the cars behind shuddered to a halt: half a mile ahead, rebels had erected a fortification that blocked access to the highway.

They were stuck, out in the open, midway between whatever cover remained in District 2 behind them and the rebel position in front of them.

For a moment, no one knew what to do. The vehicles were stuck, out there in no-man's-land.

The first Paveway exploded so close to Gaddafi's vehicle that shrapnel shot through his windows and the airbags deployed.

Around Gaddafi, vehicles peeled off in all directions. Drivers started reversing, spinning, some gunning back toward District 2, the ones with mounted weapons accelerating toward the rebels. Men hung off the bouncing trucks and fired as they bore down on the rebels; other men leaned out of windows and launched RPGs.

Gaddafi's driver pulled out at an angle, not toward the highway ahead or the district they'd just left but across a stretch of flat land; his guards in cars nearby saw and followed, and now Gaddafi was in a spontaneous formation bouncing across the sand, racing for cover, separating itself from the rest of the convoy and looking, from above, like a cluster of vehicles that must be protecting someone important.

Two more Paveways struck simultaneously, more shrapnel tore through the cabins, and now there was fire and noise all around Gaddafi.

He swung his door open. Jets roared by overhead. He started to run. Guards in one of the cars opened their doors to follow, but only a few made it more than a few steps before flames reached the cache of hidden ammunition, triggering a series of secondary explosions that knocked them off their feet and incinerated the passengers who didn't get out in time.

Gaddafi kept running, the survivors alongside him, while behind them, around them, the escape route became a half-mile-long battle space. Air strikes destroyed vehicles and shunted surviving passengers into gun battles. People ran for cover in a nearby power relay station, a water treatment plant, a compound of walled villas surrounded by cement blocks and construction materials. Within seconds there was wreckage and bodies everywhere. Gaddafi ran toward one of the villas and hid inside, several of his men on his heels, but now it was clear to the rebels what had been clear to the aircraft above—that someone important was separating from the herd.

Almost immediately the villa started taking heavy fire, antiaircraft rounds and mortars detonating nearby, and Gaddafi's men knew they had to move again. They ran back out of the villa, reached a series of construction blocks, the closest thing to cover, and dove. The small circle of Gaddafi's men still with him were frantic. They'd been spotted, they were running out of time. They had no good options. One of them spotted a concrete drainage pipe on the other side of a road. It was a hundred yards away at least, over flat exposed sand, but staying still was worse.

They stood and made one last run for it. Bullets zinged by as they scrambled up to the road, across it, and down the other side, dropping down into the pipe, somehow unscathed.

There was a moment of reprieve, but now there were rebels coming toward them on foot.

They needed more time. One of Gaddafi's guards started pulling pins from grenades and throwing them out of the pipe, but one caught the lip, bounced, and fell back into the pipe. The guard grabbed for it and reared back to throw it again, but he moved too slowly. The air in the pipe split open and light vanished from the space.

Smoke began to clear, and as the darkness lifted, Gaddafi was bleeding from a gash in his head, the guard from a mangled arm; one of the others was on the ground. Gaddafi was in a daze, and soon rebels were dropping into the pipe, arms wrapping around him, wrestling him out into the open.

The moment was confused; the rebels fiery and violent even though they still didn't seem to know who he was until a hand grabbed a fistful of his hair to yank his head back and sunlight hit his face. Someone yelled, "Muammar!"

He managed a question. "What's this?" he said. "What's this, my sons? What are you doing?"

WITHIN HOURS THE news was confirmed. Grainy cell phone footage of Gaddafi's last moments pinged around the world: after decades of oppression, Gaddafi was dead.

Operation Unified Protector was immediately hailed as the newest example of what war could be now. No protracted foreign occupation required, no foreign troops caught in a morass like they'd been in Iraq and Afghanistan. An air campaign as effective as the one against Slobodan Milošević, only this time it stopped the dictator from carrying out another massacre before it could happen.

And this time, America hadn't taken charge. The forces of good could serve the world even without its one superpower having to lead them: it was clear now that others could take the lead in policing the world.

The United Kingdom had led a real international coalition, in which no single NATO member had borne an outsized financial burden, or suffered a disproportionate human cost—or *any* human cost, for that matter. NATO hadn't suffered a single fatality, or even a significant injury.

Heads of state celebrated NATO publicly. The organization celebrated itself. The U.S. Permanent Representative to NATO and the NATO Supreme Allied Commander co-wrote a *New York Times* editorial to make sure the general public understood what had just been accomplished. "As Operation Unified Protector comes to a close, the alliance and its partners can look back at an extraordinary job, well done," they said. "Most of all, they can see in the gratitude of the Libyan people that the use of limited force—precisely applied—can effect real, positive political change."

In Libya, in the immediate aftermath, the rebels shared a video of David Cameron standing in a just-liberated city, having flown in to congratulate and thank them for proving what was possible. It was like the British prime minister was speaking right to them. He stood at a podium and told them he hoped the world might "see democracy advance in other countries too. I believe you have the opportunity," he said, "to give an example to others about what taking back your country can mean."

The French president took the mic next and went even further in his praise of Libya's rebels as heroes to people beyond just Libya, saying he thought "about the hope that one day young Syrians will be given the opportunity that young Libyans have now been given" and then dedicating his visit "to those who hope that Syria can one day also be a free country."

The rebels had suffered losses but they now had reason for hope in their country, and pride in how they might have inspired people elsewhere. So as protests in Syria tipped toward rebellion as well, it felt only natural that NATO would shift its attention there. With Gaddafi gone and the rebels now in control, NATO considered its mission in Libya complete.

Less than two weeks after Gaddafi's death, Operation Unified Protector officially ended. The future of Libya, and the future of precision airpower, seemed for the moment to be secure.

8.

Manchester, England

The imams caught on to it early.

Even amid a celebration shared around the world for a hopeful Arab Spring and the surgical war in Libya that had supported it, there was a problem.

Young men were returning to South Manchester having seen things they should not have seen. They were returning with nightmares and drug addiction; they were like the troubled veterans in the movies, except some of them were kids.

Salman was still sixteen and reentering life after a total immersion in something both heady and awful, an experience to run away from and bury under substances but also to yearn for and try to claw back.

He looked for diversions. He partied, he went to raves, he went to the mosque five times a day. He found a South Manchester source for tramadol, the battlefield pill that switched the mix of chemicals in his brain back to what it had been in Tripoli and jacked him up to fight regime forces that were no longer there.

He tried to reenter school. He'd freed a country; passing sixth form should've been a cakewalk. His tramadol use got worse. His mother grew so concerned, she went to the family doctor for help, but when Salman tried to quit, the withdrawal symptoms were extreme.

He got dizzy and weak; he thought a little about suicide. In the revolution there had been brotherhood and glory and there had been the casual disassembly of human bodies. He started using again.

He tried visiting Abdalraouf, took him out, and pushed him around in his wheelchair, but that was confusing too. Abdalraouf was now an invalid who needed help cleaning himself, a proud lion rendered pathetic by a sniper he hadn't even seen. But instead of souring Abdalraouf on the whole idea of fighting oppressors, it seemed to make him more committed. He had the prime minister's speech on his mind, how David Cameron said the fight "goes beyond Libya," the idea that Syria needed help too. Abdalraouf had it in his head that he could do something. He knew he was no use on the ground, so he was trying to be a kind of travel agent, helping friends and relatives who wanted to go fight another dictator. But the fight in Syria was already dark and confusing. NATO was carrying out air strikes there, but they weren't bombing the dictator; they were bombing other people who were fighting the dictator. It seemed NATO was less offended by Bashar Al-Assad killing civilians than by the Islamist organization operating there.

When Abdalraouf was arrested, it wasn't hard to imagine why. He'd probably run afoul of some legal technicality and been picked up by police trying to scare him a little. But he was an invalid, and all he'd done was what Prime Minister Cameron had asked them to. Surely they wouldn't hold him for long.

Then Libya started to get dark and confusing too. Salman watched the country change. Gaddafi's death started to look less like the end of a conflict and more like the beginning of one.

If what they'd been through hadn't actually freed Libya, what had it been for? For what purpose had his friend lost the use of his legs?

Salman found more trouble. He went joyriding even though he didn't have a license. He got in fights. He punched a woman; he grabbed tracksuit pants off the rack at a sporting goods store and ran. He got high and looked for rules to break.

Imams from local mosques worried about the behavior they were seeing in some of the young men who'd fought in Libya. They seemed to have come back changed.

Libya held elections; the results were disputed. The people were splitting apart. The country careened toward more violence and other governments no longer cared. Salman saw the prize they'd sacrificed for evaporating. He decided to go back.

9.

Tripoli, Libya

The Libya that Salman returned to was a disaster. The revolution he'd been a part of—the revolution that had made him an addict and made Abdalraouf a paraplegic—was now an orderless war.

The West was nowhere to be seen. Embassies were shut, staffs evacuated, and after bombing the country for five months, NATO had abandoned it. Now it looked less like they'd freed Libya and more like they'd just broken it and then left to bomb people somewhere else. The country was crumbling and no one cared. There was now constant fighting and kidnappings; there was no functioning government because there were two competing ones and neither was able to maintain any kind of order. Militias emerged, and not all the people in them were Libyan.

As Salman tried to make sense of the unstable terrain, he began to encounter some of the foreigners. The most impressive were people who said they belonged to a group setting up a base of operations here because NATO was bombing them in Iraq and Syria; they'd needed a new safe haven, and they'd found it here, in the chaos crippling Libya. They said their leader was no less than a caliph. His name was Abu Bakr al-Baghdadi. Their organization was called ad-Dawlah al-Islāmiyah: the Islamic State.

They found the lawless helpful, and the lawlessness seemed to be the worst in the places the air strikes were most concentrated. The city of Sirte, where Paveways helped kill Gaddafi, was now in anarchy, so the Islamic State concentrated its operations there. But if Libya was in chaos,

Salman found the Islamic State fighters moving in were disciplined. Two bickering Libyan governments did nothing while the Islamic State guys were committed, they had purpose and organization, a stark contrast to everything else. Even more impressive because many of them were part of a particular Islamic State brigade, an elite commando unit al-Baghdadi had formed as part of a push to make his "state" capable not just of defense but offense, too: he wanted to fight enemies where the enemies were, not just stop them when they came to his territory. So he'd composed the special commando unit of recruits from all over. Its membership was diverse, it included people even from Western countries. They spoke different languages, *European* languages; they were something like the Islamic State's external operations bureau; they were looking to send fighters in Libya to launch operations in Paris, in Brussels, in other cosmopolitan cities. As a kid with a British passport, fluent in English, born in the UK but with increasingly complicated feelings toward it, Salman was exactly the kind of candidate the unit was trying to recruit.

WHEN SALMAN RETURNED to England, friends noticed more changes. He spoke with more gravity. Things he'd taken seriously before he now treated with indifference. He got in arguments, he was more stern, but he also appeared to be more in control. He seemed less hotheaded: no more petty theft. Somehow, outwardly at least, he appeared to be off the drugs, and to be taking responsibility for his life. Friends said he must be finally listening to his mother; she was probably telling him to shape up so he could find a wife. He wasn't partying. Friends welcomed the changes. It looked like maturity.

To other young Libyans in Manchester, though, Salman had an edge. He started railing about all the people being bombed by Western planes. The whole world grieved every time an American or European was killed, but the thousands of people killed in Iraq or Syria or even Libya barely registered. He tried to convince people around him to be more serious. "Life's temporary," he told a mystified friend, "make sure you don't get lost."

And at the mosque, worshippers were getting worried. After an imam gave a sermon condemning the Islamic State, Salman approached, said nothing, just glared with obvious menace. Members of the congregation called a counterterror hotline. They didn't know what was going on; they couldn't know for sure; the idea that Salman may not have entirely left behind whatever he'd gotten up to on his travels was just a suspicion. Because when he actually began communicating with members of the Islamic State commando unit he'd met, he was careful to avoid detection. He was tight-lipped about it even with Abdalraouf, who was out on bail now and about to start his trial, but was perhaps being monitored by authorities.

Salman used burner phones. He didn't talk to the Islamic State contacts directly but through an intermediary in Germany who forwarded text messages back and forth between Libya and Manchester.

And if Salman's commitment to the Islamic State was not yet absolute, in the spring of 2016, he received three pieces of devastating news, each almost perfectly calibrated to push him over the edge.

He learned that a friend of his—another Libyan kid from Manchester who'd left to fight—had been killed in a precision air strike. Surely by Western planes.

Then Abdalraouf's trial ended with a guilty verdict and a nine-year sentence. Now Salman was watching a close friend who'd taken bullets for Britain being locked away for doing what he thought Britain wanted him to. As if he were an inconvenient reminder of something they were maybe ashamed of now. Locking up a cripple to hide their mistakes.

The day after Abdalraouf was sentenced, Salman learned that another young Libyan friend had died, this time right in Manchester, and the details were gruesome. Salman's friend had been attacked, managed to get up and run, only for his attackers to chase him for more than five miles in broad daylight, before hitting him with a car, attacking him with a hammer, knifing him in the neck, and leaving him to die.

Salman felt he saw the murder for what it really was: a religious hate crime. The police dismissed it as gang-related, which made it even harder to swallow—a way to make the victim look more like a criminal who probably deserved what he got. Another expendable Libyan kid no one would miss.

Salman had a harder and harder time hiding his anger toward the British government. This country that had people like him go fight and die and be paralyzed when it was convenient and then imprisoned them for it. A country that didn't care if people like him were hunted down in the street.

Salman took more concrete steps. He opened a new bank account. He convinced his younger brother to help him call friends and get them to make a series of odd purchases. Each one came with a plausible explanation: ten liters of sulfuric acid from a supplier in Italy to fix a car; another five liters for a generator battery in Libya; thirty liters of hydrogen peroxide from a Scottish bottling company catering to the e-cigarette industry; more than sixty-five pounds of nails and screws from three different hardware stores. His movements took on new urgency. He bought a hatchback, rented apartments around the city, and separated his activities between them so it would be harder for authorities to connect the dots. He had the chemicals shipped to an unoccupied flat in one neighborhood, the other equipment shipped to a different one across town. The strategy worked: no one seemed to know what he was up to, but it made the planning harder because he had so much space to cover and not much time. Now that he was acquiring materials, the risk of being found out was greater and the window of opportunity to get it all done began to close. He rushed. He raced back and forth between the apartments, distracted by the task at hand and not paying enough attention to the road. Driving too fast, picking up boxes and shuttling them from one apartment to the other to assemble the materials before it was too late, nearly undoing the entire operation on one final supply run when he and his brother ran a stop sign near a girls' school in a Manchester suburb and struck another car with such force that it sheared the hatchback's bumper off and the radiator shot off its support. They only barely got away before police arrived, but now there was even more urgency.

At one of the apartments, just northwest of Manchester, Salman began preparing the explosives, a stage that shortened the time he could expect to go unnoticed even more. As he mixed the chemicals in a blue industrial drum he'd stored there, the sulfuric acid gave off a vapor strong enough to sting nostrils, and the acetone made the halls outside his unit smell like fruit. He finished the mixture before anyone was able to identify which apartment the smell was coming from, poured the chemical compound into smaller containers, disabled the fire alarm, and followed instructions for assembling the electronic components—the final step—of a seventy-pound bomb.

Salman scrawled a note on a paper towel for the landlord. "Sorry if there is anything missing in the apartment or any mess made the extra 150 and the rest of the week's rent can hopefully cover it."

And then he left.

10.

Eastbound on Route I-94

Kathy Kelly was packed into a bus full of people headed for Canada like some kind of draft dodger making a run at the border.

She didn't like flying so much anymore. It wasn't a phobia; it was conscience. She'd begun to think about all the places her work had taken her. She was beginning to think she might be responsible for as much jet fuel consumption as the Platinum status guys, the million-frequent-flyer-mile businessmen.

But at sixty-four years old, she wasn't ready to stop traveling. So she started going by road.

She was by then a Nobel Peace Prize nominee with bad hips and bad osteoporosis taking long Greyhound rides that felt like being back on the inmate transport to the Lexington federal. But there were still trips she needed to make.

Now, she heads to a Peace Camp one of her radical friends is involved in. A good friend, from one of the Christian Peace Teams, and he was once held hostage in Iraq for 118 days; she knows the number exactly. She feels loyal to him, and even if she's never been the most puritan in her faith, lately she's been feeling the need to repent for all the carbon, so she's punishing herself with nine hours on a bus. Anyway, it's a camp for teens, for youth, and *kids are always innocent*, and the camp is dedicated to helping the teens work for peace, so there are six different reasons she couldn't say no.

Lieutenant Rick Hilton was living in San Antonio, beginning to lose touch with the friends he fought with. For a while he and Ann wintered out in Vegas, where Hilton golfed with guys who'd been on active duty at the Air Force base just outside the city; where he saw pilots from his old squadron. Where he kept up.

These days, he's having a harder time. Mostly it's the distance between Las Vegas and San Antonio. Just a thousand or so miles, barely an hour in a Phantom at top speed, and it's a little strange how a distance that feels like a universe at home in his own country would have been trivial in war.

Here, the distance is enough to ease him out of touch. The distance and the other thing: there were pilots who looked immortal for making it through the onslaught of ground fire at the Dragon's Jaw, but time is a trickier thing to dodge. Pilots who survived Vietnam were starting to die of more pedestrian things; it was happening too fast and Hilton was more and more disconnected. Some days he found himself sunk into a chair, unable even to move. Pinned in place by the rising and falling body of a yellow Lab that's taken to sitting on his feet.

She'll be gone soon too. The latest in a series of dogs he and Ann have been raising, the consequence of a different attempt to stay connected. They'd looked into raising puppies destined for military training programs but found the house they were living in and their age were poorly suited for the high-energy breeds favored by the military, so they settled for the closest program they could find: raising puppies until they were old enough to be trained as guide dogs for the blind.

And this was a strange thing too—that of all people, it would be him, an ace capable of rapid math at altitude, a liquid-cooled brain making life-and-death decisions at high-Mach speed, now occupying himself with this gentler pastime. Perhaps another casualty of later life. Some age-related atrophy of the combat-honed muscle that kept him hyper-rational, that used to let him replace emotion and attachment with logic and odds.

Or perhaps it's the opposite. Perhaps it was that same talent, strong as ever. How else can you repeat this routine, having a creature dependent on you just long enough for you to grow attached to it before you have to let it go, again and again? He knows each time another new eight-week-old stumbly little thing is handed to him, to crash into furniture and pee on things and then sleep for twenty hours a day, that it's best not to get too invested. And still, the pride takes him by surprise when he sees them grow up and out to lead strangers into some unexpected version of their own future.

Not far from where Rick Hilton sat, Weldon Word had settled into his own retirement, living just off a golf course an hour or so from Dallas, having mostly lost connection with the weapon he'd created even as that weapon was out in the world, realizing his most optimistic ambitions for it. By then, after ramping up to meet rising global demand, both companies building his creation were seeing record sales. Lockheed Martin had just reached its peak of over 120 produced a day and was about to surpass 100,000 delivered since it first joined Raytheon in manufacturing Paveways in 2001.

Raytheon, for its part, was seeing such significant growth in Paveway sales that even as it sold a range of other weapons systems—more elaborate ones, and ones for which it was the sole provider—the company declared Paveway to be the "weapon of choice" for what was now more than forty-three customers all over the world. Paveway was helping to drive Raytheon's biggest revenue quarter, and its biggest revenue year, bringing in more money than any other product in the company's missile systems division. Even though Paveway remained, by far, the division's least expensive product.

And while Weldon spent his days volunteering at church and finding smaller problems to fix around the house, the work he'd begun on Paveway was enabling and inspiring new capabilities that would've been hard even for him to have fathomed. The U.S. Air Force was working to outfit one of the autonomous drones that carry Paveways with a supercomputer designed to identify targets on its own, as well as track them, evaluate them, and "nominate" them for engagement. The work had astonishing implications. The Defense Department seemed to anticipate them, and issued a policy directive pledging there would always be "appropriate levels of human judgement over the use of force." But the capabilities even of that one program, and the fact that the drone was designed to be able to operate even if it lost communication links—to operate, in other words, even if a human wasn't operating it—betrayed the growing temptation to take the fateful next step in a process of evolution that began with the first Paveways. A step not just toward making it cheaper and easier to fight; not just toward allowing us to fight wars while removed from the danger; but toward allowing us to fight wars while removed even from the decision to attack.

11.

Manchester Arena
Victoria Station Approach
Manchester, UK

On the evening of May 22, 2017, more than 14,000 people assembled at the biggest indoor arena in England, a world-renowned, American-owned venue built onto one of the country's biggest railway exchanges.

British authorities had issued a "severe terror threat level" because of increasing chatter about young men returning from conflicts abroad and because the Islamic State had begun carrying out major terror attacks in Europe. They'd struck Paris, then Brussels, and claimed credit for an attack a month before in Westminster.

Three different organizations provided security for the Manchester Arena, but, due to a typo, the arena's security contractor had distributed a risk assessment mistakenly identifying the terror threat level as "low" rather than "severe." The security perimeter had not been extended beyond the City Room, an entrance hall with unclear jurisdiction between the police, the concert operator, and the private security contractor. Access to the City Room was not controlled. Members of the public entered unsearched.

Salman Abedi walked into the City Room at 8:51 p.m. after arriving at the connected tram station, walking past two police community support officers on the station concourse, taking an elevator up to the arena level, and passing two security guards. None of them questioned him, despite the fact that he was overdressed for the warm night, visibly weighted down by an apparently heavy backpack, and that he didn't fit the audience demographic for the evening's act: an Ariana Grande Dangerous Woman Tour performance.

Once inside the City Room, Salman carried his backpack up a short staircase to the mezzanine.

Parts of the mezzanine were visible on screens in a control room showing CCTV feeds from dozens of cameras all over the arena complex, providing a sense of near-total control over the space, but Salman found a corner where no camera pointed. He hid and waited.

At 10:09 p.m., with the concert scheduled to end in twenty minutes, a security guard began a pre-egress check, walking through the City Room to make sure the path was free of obstructions for the thousands of exiting fans who would soon fill the space. Because there were no doors to the concert space up on the mezzanine, the guard did not climb the stairs and gave the mezzanine only a passing glance. The guard did not see the figure hiding there with a backpack.

Ariana Grande began her final song. There were no more security measures left to stop Salman.

At 10:22 p.m., with the concert scheduled to end in only eight minutes, Salman caught the eye of a man waiting to pick up his girlfriend's kid. It looked like Salman was trying to hide. The man saw Salman's backpack. He wondered if it was possible that the backpack might contain something like a bomb.

Through the doors, Ariana Grande finished her last song.

The man went to find a guard.

Ariana Grande came back out for an encore: her album's title track, "Dangerous Woman," which typically lasted around four minutes.

It took almost five minutes to find a guard with a radio. When the guard looked up toward the mezzanine, Salman got fidgety; he looked nervous. The guard thought it suspicious enough to alert staff in the control room, who could notify an event manager, who, in turn could have the City Room sealed off and direct the crowd toward the other three exits.

But as Ariana Grande finished her encore, and people began to move toward the exits, the radio channel grew more crowded with staff coordinating the movements of 14,000 people leaving the concert hall. The guard trying to report a suspicious man hiding on the mezzanine couldn't get through.

The City Room began to fill. Children darted around looking for parents; families tried to find each other.

Salman got up, walked down the mezzanine stairs, and moved across the floor into the middle of a squeezing-together crowd.

The guard tried the radio again. Bodies pressed against bodies.

At 10:31 p.m., Salman detonated his bomb.

12.

Royal Air Force Akrotiri
Western Sovereign Base Area, Cyprus

Three days after one of the worst ever terrorist attacks on British soil, a photograph began to circulate on social media: on the side of what looked like some kind of guided weapon, someone had scrawled a handwritten message:

WITH LOVE FROM MANCHESTER

There was initial skepticism, but the British military quickly confirmed it was real. The weapon was a new Paveway IV, mounted under the wing of a plane about to take off for a new mission against new foreign targets.

Afterword

ELKAN

Sarasota, Florida

The transition was difficult. Florida was unlike any place Radica had been.

No one in her family spoke English. It was hard to find work. Eventually her husband got a job hanging storm shutters. Not exactly using his engineering degree to its full potential, but it turned out shutters were a growth industry in Florida, where everyone was always thinking about hurricanes, and Radica's family had their own experience with wind and shattering glass. No one needed to know from where.

Radica tried to keep an open mind. She told herself she was willing to take any job so long as it didn't have to do with old people, which seemed to be what everyone did here. Old people disgusted her. Having to change them, clean dribble from their lips, smell their old-people smells. She wanted nothing to do with the frailties of human bodies. She'd seen enough of that back home.

But the only other Serbian in Sarasota worked with the elderly, and the only job she could find without speaking the language was serving food at an assisted-living facility, so she decided that was acceptable, the kitchen far enough away from the rooms with all the decay. Anyway, she was distracted by what Kiki started going through.

Kiki started her new American school and promptly began drowning in it. Kiki cried for a year. Her math placed her two grades ahead, but anything to do with reading, speaking, understanding a language she'd never spoken, was hopeless. She couldn't do her homework. She couldn't even start her homework. Radica tried to help, mother and daughter huddled over a kitchen table with their Serbian–English dictionary, translating homework instructions one word at a time. Radica, near tears herself, dragged a translator to Kiki's school. "I want to help her but I don't know how!"

The teacher said something. The translator translated: "Don't worry. You'll see. In six months she'll be the best student here." This American brand of fantasy. Radica couldn't just watch the girl flounder. What kind of mother couldn't even help her children with homework? Radica needed to learn English. She'd have to go behind enemy lines. She put in for a different job at the assisted-living facility. She braced herself and moved from the kitchen to the dining room. Hostile territory, the elderly all around, but she'd hear more of the language. Her boss gave her a supervisor's name tag, put her in charge of eggs, saw her straining to understand what was happening around her, and started giving her his newspaper every day, somehow making clear that she was not to come back from her lunch break until she finished it. And when he offered her another change, from supervisor of eggs to a less glamorous overnight job, she decided to take it, because the old people would be asleep and she wouldn't have to touch them much, but mostly because the night shift meant she could help Kiki during the day, and all she had to give up was sleep.

Kiki flailed. They put her in an extra English class next to a Denny's and she struggled through that too. The move from a small town in Serbia to a huge country shrank her world. There was no more wandering. People didn't even like to walk here. Kiki sat in her strip-mall English class with the rest of the misfit immigrants, looking out across the concrete, daydreaming about Lučani, wondering what she'd done to deserve this punishment. Her mind wandered, her gaze shifted, her eyes fell to a hand on a desk next to her, guarding a notebook with a photograph she recognized, an airbrushed soap opera star. Kiki heard herself yell in Serbian: "I know that! I know that!"

Rebelde characters on a notebook. Kiki hadn't known such a thing existed.

The teacher stopped talking. Kiki thought of her grandmother. The notebook's owner looked at her. Neither could speak to each other, but soon the Serbian girl and the one who turned out to be Venezuelan were bonding across their few shared words and the beautiful *Rebels* glossed in fake tears. Things shifted immediately: now Kiki had a friend, and soon they were expanding their horizons together. Huddling in front of the TV like she'd done with her grandmother and learning together that there in Florida, if you hit the UP button on the remote enough times, you crossed some invisible barrier and the televised world shifted to Spanish.

For the first time since arriving in America, she could imagine herself in Serbia. With her Venezuelan friend, a Mexican prep school on TV, and a new slate of Spanish soap operas on the higher TV channels they explored together, Kiki finally felt close to home.

A FAMILY GAINED momentum. Radica was offered still another job in the nursing home, this time as a med tech. Now supervising medicines the elderly took. Radica was handed the power to keep them alive, or at least to keep them comfortable, these people from the country that had destroyed her hometown and killed her friends. But Radica learned a shocking thing. Now that she was around old people all the time, she found she actually didn't mind it so much. Most of them were patient with her halting English. She found herself drawn especially to dementia patients. Something about the absence of pretense. It collapsed the distance. When they were happy, they smiled. When she made them feel good, she knew right away. She'd spent weeks of a bombing raid yelling into a radio to warn her neighbors, without knowing whether anyone heard, whether anyone had made it to safety.

Now, a world away, with these old slowing-down people from the country that had done that to hers, she found connection again. If she showed kindness, and a demented old lady smiled, she knew she was doing good. There was no waiting or wondering.

Radica began to notice things. Didn't it seem like all Alzheimer's patients had blue eyes?

Didn't it seem like they were all *smart*, or had been—engineers, lawyers, the decorated general she cared for who'd whistle through his teeth, dive to the ground, and shout out complicated-sounding military maneuvers? She squinted, decoding what went on inside these minds. Her English caught up with the things she was noticing. She got permission to attend a seminar where she was asked if she wanted to know what this strange affliction, *de-men-she-a*, felt like. They put some kind of insert into her shoes, an elaborate headset and gloves and earphones on her, and then gave her simple tasks to complete. "Go into that room, grab a yellow blanket, fill a glass with water, and write a letter to your parents."

She stepped toward the room and everything exploded. Sirens blared in her ears, something started flashing in front of her, weird pulses strobed on her legs and hands, and even though she knew she was in a simulation, it only took a few seconds before she was overwhelmed and crying, groping for a pen and writing to her parents, stuck somewhere back in Serbia.

"I don't know where I am, I don't know what to do," and then she was shaking too much and heaving too hard to complete the rest of the exercise, but she knew this was her calling.

Her world became the space of breaking-down bodies, days spent tending to their infirmities. Enduring their smells, changing their diapers, learning to love them just as they began to waste away. She worked in that opiate haze, among murmuring waves of family members with their elbows in their palms and fingers on their necks, urine-soaked fabric that needed to be cleaned but for now best to just to let things be. She became an expert at helping people lose their minds a little slower by doing all the things she thought when she landed in this country she'd never do. She massaged their backs if they liked it. She pushed them onto their sides to prevent bedsores, she learned to dodge their fists when they swatted her out of reflex. She washed them so they'd be clean when they left. She was always their friend by then.

SOON, SHE BECOMES one of those lamenters in ancient Egypt, not tearing her hair out but perhaps the modern equivalent, becoming proficient at the mechanics of death. Clever, tactful things on a dying man's iPhone that make him seem more celebrated. Absentee family members send text messages to bedsides; she commandeers phones and texts back. On behalf of a woman whose facial muscles are now paralyzed: "She really smiled when she got your message." Radica learned to plan ahead. She extracts little tidbits from the dying while she still can, and later deploys them as her own little guided weapons against guilt. "He was so proud of your girlfriend, he talked about it all the time." She means: *It's OK that you can't be here. I am here.*

She slips into the shrinking families and finds herself wanting to care for all of them. Cousins in a big city, grandchildren in the Midwest, a brother kept away with his own health concerns. Children consumed with their own indispensable Service to the World. Everyone is everywhere except for her, because she decides her job is not just comforting the dying but tending to the regret that stirs inside a family and can tip into accusation during final approach. She wants to free everyone from guilt, if they're not coming anyway. Her only weakness is that she lacks a tool that's crucial in a field like this: the power to dissociate. She becomes attached to some more than others, but she always becomes attached. She's devastated, always, when they're gone.

But she has purpose. She becomes so good at caring for these people, so passionate about it, she hangs a shingle. She gives business cards to the stylists at salons, to the clerks at the bakery, to everyone she can think of, hoping that in America it wouldn't be too hard to build a roster of old people who need memory help.

She starts hearing from people all over West Florida. After she became far too busy, she gets a call from a man who saw her card at the barber's. "This is Elkan Ries." He speaks fast. She pictures a serious man, impatient. "I need some help with my wife. Memory problems."

She tells this Elkan she's too busy already; she doesn't have any openings for new clients, but she promises to find someone who can help him. Elkan persists, and when weeks pass and she still hasn't found someone else to help, she relents. She'll somehow make time, because in spite of everything she learned in America in the years that had passed now, she still hadn't quite figured out how to let people down.

She meets with this Elkan. There's something different about him. A feeling of the popular guy giving you attention. He flirts with waitresses, a man so merciless that everyone loves him. He tells Radica he's worked out a scheme: he doesn't want his wife, Shirley, to know why Radica is there, so Radica should pretend she's a fitness instructor. Radica doesn't know the first thing about being a fitness instructor, but he says not to worry, his wife doesn't have the most ambitious fitness goals. She aspires mostly to riding the stationary bike without falling off.

So Radica fakes it. She says, "You're doing great!" while Shirley pedals for a few minutes, then the two go out for lunch. Radica begins to learn about Shirley: the store she once owned, the three daughters she had with Elkan, her strange habit of sneezing after a nice meal, how that drives Elkan crazy.

Radica begins to see, now that she's become so proficient at seeing past words, that Shirley regards her Elkan as something like a god. She loves this man who no longer has patience for her.

IT MAY BE that he's losing patience with everyone. It seems at first he's losing patience with Radica too.

"We went to the store," Radica reports to Elkan, after an outing with Shirley. "We had *wery* nice time."

"Speak English!"

"We vent to a store *vare* they had flowers she likes."

"'*Went*'! *Where* they had flowers! Speak *English*!" Elkan means to strangle the accent out of her. Elkan promises he'll teach her to speak right.

Radica promises back: "I'm *newwer* going to lose my accent; that's not how accents work."

Elkan tells her, "People won't treat you the way you deserve to be treated if you sound like a foreigner." Radica tells him, "You don't treat your wife the way she deserves to be treated."

Elkan throws a hand as if to dismiss her, but he may be hiding a smile. He's impressed.

He begins to lay traps in conversations for her. He baits her into saying *w* words—"when," "where"—so he can pounce when they came out "ven," "vare." He assembles explanations of important things around the longest words he can find.

"Elkan, I don't understand this word!"

Sometimes he gives her the definition, sometimes he snaps, "Get a dictionary!"

She comes to the apartment one day to find he's bought her books on America's founders to give to her children, the same books he buys for his own grandkids. He teases her, he loves her, he loves his wife, he just needs Radica to help him see it.

Eventually she sees Elkan's way of showing affection is ridicule and that Shirley gives it back as much as she takes it. That, or some instinct in Radica registers this as one of the white lies she can tell their relatives, if she ever has to meet them, so that when the time comes, they won't have to remember Elkan as a bully and Shirley as a victim.

When Shirley's memory becomes too porous to handle the tasks of daily living even with Radica at her elbow, they decide it's time to move her into a nursing home. During visits, conversations between husband and wife shrink into an ever-tighter loop, and Elkan's already stunted patience drains even quicker. Radica's job, never formalized, expands to preventing Elkan from shoving his wife off a balcony—or jumping off one himself.

Elkan endures depressing holidays in the dimly lit cafeteria. They have a quiet nursing home Thanksgiving, just the three of them, sitting on the laminate chairs, passing Styrofoam back and forth. Radica knows Americans care about this holiday, so she's brought takeout.

"El," Shirley says, "have you spoken to the girls?"

"Yes, they're fine," Elkan says. Already a little too short, Radica thinks.

"Radica, hon," Shirley says, "can we heat the stuffing up? It's a little cold." Then: "El, have you spoken to the girls?"

"Damnit, Shirl, you just asked that!"

"Oh, shut up, Elkan! Can you just—"

Radica interrupts to yell something amazing about the weather.

All three are quiet for a moment. Then Radica laughs, and Elkan and Shirley laugh, too, each aware of exactly what Radica is doing, laughing because it's so obvious, and because even still it works. The impatient old man, the demented old lady, the caretaker who's already more than that, laughing until their eyes get a little wet over their disposable plates and their sad little happy Thanksgiving dinner.

"I'M LEAVING FOR three weeks, Elkan," Radica says one day. "Ve're going back to Serbia." She worries for them. Elkan is too impatient around Shirley without Radica there to enforce the ceasefire. Elkan is too proud to take care of himself. "Are you going to be OK ven I'm gone?"

Elkan looks stricken. But his face changes. He hollers, "*When* you're gone! Speak English!"

She goes home to see her own parents, a too-short trip, and when she comes back, Shirley's taken a turn. Radica goes to the hospital and can tell immediately that Shirley's terrified. She's woken up with half her life gone. The bank of faces in her brain wiped almost clean, and Radica can see on her face that everyone is a stranger. Now she's under siege by an occupational therapist who seems intent on proving how little the patient has left. "Honey, do you know what year it is?" Shirley looks away. Didn't Shirley use to read three newspapers a day? "Do you know who the president is? . . . Do you know what state we're in? . . . Do you know what month it is?"

Radica feels her chest tighten. She's seen this before. She's seen people lose their ability even to speak. She watches Shirley blink around the room, defeated, retreating.

Then, her sapphire eyes lock on Radica, and Shirley turns back to the therapist.

"That's Radica. Did you know Radica is from Serbia?"

Radica feels the chambers of her heart bang together.

ELKAN VISITS HIS wife as often as he can bear, but it's Radica who steps in and becomes their relationship, holding on to the last tattered fibers connecting them. Running from Shirley at the nursing home to Elkan and back, sifting through the words they have for one another, relaying only the kindest ones. Radica can see Elkan entering a trap that she'll have little power to get him out of. Being with his wife is impossible for him, but being away from her is harder than he expected. He was proud of the fact that in his retirement he didn't feel the need for a legion of friends, but now the few friends he's kept seem to suddenly all be dead or unable to keep up, his family is scattered all over the country, and his wife is in an intolerable place that reminds him only of decay, installed across town in her own shrinking world.

Radica gets a call. "It's me," Elkan said. "I'd like you to come to take care of bills and to give me updates on Shirley. Maybe once a week."

There are no bills to take care of. She already gives him regular updates on Shirley, but Radica understands. It isn't about any of that. It's hard for Elkan to explain, hard to quite put his finger on it, but he'll wake up at odd hours and wander around his little Florida neighborhood, find himself sitting on a public bench. He'll call his younger brother, both of them around ninety now.

"I'm sitting on that bench near the church."

His brother smiles into the phone. "You're converting?"

Elkan sits for a minute. "I'm lonely."

They haven't talked much in their later years, but Cary, never all that close, knows what Elkan is saying.

Elkan is giving up. He starts sending things to people. Money as gifts, small investments in businesses he knows won't succeed. But he invests anyway because it's a silent way to give. Detaching himself from possessions, as if they might hold him back. But a person has come into his life from a splitting-apart country ready to ease him on. A woman coming from a home cracked apart by American bombs, to a man whose home has been cracked apart by simpler things, age and the atomization of family, the simple nature of decay.

He starts falling. Radica gets troubling phone calls.

"I'm on the floor."

"My God, Elkan, you call 911."

"No."

"Elkan, I'm in Englewood. It'll take me hour to get there."

"I'll wait for you."

Two days later he falls again, this time in the middle of the night. Too far to reach the phone, and he doesn't want to press the button on the Life Alert he now wears around his neck. He stays on the floor all night, waiting for Radica to find him in the morning.

"If I push the button," he says when she finally shows up, "I go to the hospital. If I go to the hospital . . . I don't come back."

She's irate about this stubborn man, elbows bloodied, an idiot on the floor.

He's almost ready but not quite ready yet. He doesn't quite understand what's holding him back until he's fallen for the final time. Radica helps make arrangements for a hospital-style bed in his apartment. She helps him ease into it. He wants to speak to Shirley, to see her face, but he can't bear the thought of her seeing him like this. Gaunt, unshaven, shameful growths on his face. She'd be confused; she'd be horrified. It would ruin her. He misses her, he wants her, but he refuses to call to her. He stops eating. He understands now that what he's going to have to do is to leave her, forever, without saying goodbye. He understands now that they've already spoken for the last time. What had they talked about? What spiteful thing had he said?

He stops drinking water. He winces and groans through the night, but he holds on. He eats again, just for an afternoon, when Radica brings a donut to share. He loses the strength to speak except as a growl. Now that death is a real, close thing—morphine and lorazepam on a table in the next room, his bed lined up so he can see the sun setting on the bay, low music playing from an iPad wedged into a seat cushion—Radica can see the fight happening within this barely moving man. She sees the realization he's coming to. He's come to feel he's let Shirley down. The feeling of air moving through his throat makes it too painful to say. He holds on only because he worries that once he's gone, Shirley will be abandoned. She'll be there forever, wondering where he's gone, what she's done. He holds on as if by lying in pain long enough, some solution will occur to him. Or maybe he holds on as penance for what he thinks he's done to her.

Radica sees what's happening inside him. She reads the little grunts, the flick of the eyelids. His brittle hand, barely raised, barely pointing at a gaudy little picture frame on the credenza. She hears what he's trying to say. She knows what she has to do.

I WENT TO see my grandfather just before he was gone. When I got to his apartment, Radica was at his bedside, instructing me gently to leave, though I'd only just arrived. She whispered to me, "Please, I think you now go see your Nana." Giving me a kind smile and a half nod, as if to say *It's OK, I'm here.*

She gave me a minute with him first. I told him he was lucky, this seemed like a very special woman. He made a noise that was almost a word. I leaned closer, into his smell. With heroic effort he manages to get a sentence out.

"She washed my balls."

Radica, still walking away, rolls her eyes. She's heard worse.

"Well," I say, "you still got it."

I can't tell if he smiles, but his eyes follow Radica. Once she's out of earshot, he manages something else. "She's superwoman."

Radica stays outside, pretending everything is OK. Leaving me to try to make him smile, or perhaps to make myself smile. "Pop, you look terrible."

But Radica, standing out in the hall, knew what needed to be done. And when I left, when I took my grandfather's car—a car he'd never drive again, with tennis rackets in the back—and arrived at my grandmother's room at the nursing home, I saw that Radica had arranged for a friend of hers, another Serbian woman, to be waiting to take a photo of my grandmother and me smiling together. This Serbian friend sent the photo back to Radica, who held the phone up in front of my grandfather. "Look," Radica said. "They're having lunch together. *Look*, Elkan. She *von't* be abandoned."

Radica told me he smiled.

Two days later, he was gone.

Two months after that, Shirley was too.

Acknowledgments

Thank you to Margot, Jenna, Dad, and Mom, for so ruthlessly having my back.

Thank you to the Pulitzer Center on Crisis Reporting, for helping to launch my career, and for supporting the original story that led to this book.

Thank you to Willy Staley, Jake Silverstein, and *The New York Times Magazine* for turning a wild idea into a piece that maybe did a bit of what we hoped it would.

Thank you, John Parsley and David Howe, for your wisdom, patience, and judiciously applied firmness, and for doing a passable job of convincing me you don't regret all the rope you gave me.

Thank you to all the other editors. Thank you to Richard Hart, who is probably as responsible as anyone for this journey I've been on. Thank you to Cullen Murphy. Editor, friend, oracle, guide in life and story.

To Ben Adams, for your vision, artistry, gentleness, and foresight, and for helping to conceive of this idea in the first place.

Thank you to everyone at 30 Birds. I love you all, I'm indebted to each of you, and this book only exists because of a thousand ways you covered for me. Without ever once asking when it was your turn. Mohammad, Manizha, Abuzar, Tahera, Christa, Jennifer, Justin, and even you, Bella. Thank you a thousand times. To Simin and Aziz, whose mix of courage and humor infected all of us.

To Amman and Missy, who take me out into the world during my loneliest moments, get me away from my words, and remind me there is humor and community and oyster shooters to be had in the pursuit of stories.

Thank you, Amy Madsen, a secret weapon, on whose insight, heart, conscience, and imperviousness to fatigue I've depended since she first came into my life and started machine-gunning solutions at me and everyone I care about.

Thank you, Andy Kifer. It may sound like an exaggeration to say this book would not exist without him, but it's not.

Dave Larabell, you are the world's handsomest agent, a brother, and a relentless friend. Thank you. And thank you to Michelle Weiner for your foresight and for continuing to spend an imprudent amount of your time and attention on me.

Thank you, Kathy Gilsinan, for the fields and fields of support. And for Wednesday night beers.

To Patrick Malone. For never begrudging my better sports teams and for always being there with a ridiculous meme or a critical piece of source protection advice. There is no better journalist and no more generous colleague.

Thank you to Ben Kalin for taking on a bear. If mistakes snuck through, it's despite the better part of a year's worth of fact-checking and despite his unequaled doggedness.

And thank you, finally, to the legions of people on whose work and time I drew so heavily, in this book and in every story I've tried to tell. We stand on the shoulders of giants, but we also sometimes stand on the shoulders of regular people with chores to do and mortgages to pay.

A Note on Sourcing

I have always been indebted, in every story I've told, to historians, journalists, and experts whose work enabled my own. And to people who've shared parts of their lives with me, even though it's almost never their job to do so. This book is an exception only in the size of that debt and the variety of people and works I owe it to. And while there's a full catalogue of sources to follow, a few bear special mention first.

For the story in the prologue of Joe Kennedy's mission against a Nazi weapon bunker, Alan Axelrod's *Lost Destiny: Joe Kennedy Jr. and the Doomed WWII Mission to Save London* and Hank Searls's *Lost Prince: Young Joe, the Forgotten Kennedy* are both exhaustively researched narratives, and both were indispensable.

For personal details in the story of Lieutenant Colonel Richard Hilton trying to survive the skies above Vietnam told in Book I—and for many of the technical details about aerial combat at the time—I'm indebted to Richard Hilton himself. For the hours upon hours he spent in interviews with me, the secret cockpit audio recordings he made more than sixty years ago and shared with me, and for the parts of the story he told himself in *There Are No Sundays: A Youngster from Oklahoma Finds a Home in High-Performance Jet Fighters*. I'm also indebted to Stephen Coonts and Barrett Tillman for their authoritative *Dragon's Jaw: An Epic Story of Courage and Tenacity in Vietnam*.

For the spy-vs.-spy story in Book III, I'm indebted to James Richmond for expert translations of German court proceedings, and for details about Stasi officer Rainer Wiegand in particular to John O. Koehler's *Stasi: The Untold History of the East German Secret Police*.

For the story of a Kathy Kelly, a woman finding her place in the world of radical activism in Books IV, V, and VI, I'm indebted to Kathy Kelly herself for the patience with which she recounted her story to me over our years of interviews, and for the care, humanity, and talent with which she writes about these experiences herself in articles and in *Other Lands Have Dreams: From Baghdad to Pekin Prison*.

Of all the works on the evolution of airpower I consulted, two authors were especially helpful both in my own education on the subject, and in providing the details necessary to build a book around it. Paul G. Gillespie, for the time he spent on the phone and answering emails from me, and for his authoritative *Weapons of Choice: The Development of Precision Guided Munitions*; and Peter deLeon, for his 1974 report for RAND Corporation's Air Force Project, "The Laser-Guided Bomb: Case History of a Development."

And for the story of a single extraordinary invention that few knew was changing the world, I'm indebted to a brilliant, patient, and generous Texas Instruments veteran named Steve Roemerman.

Thank you to those authors and experts, thank you to everyone else who lent their time and expertise to this journey, and thank you to *you*, the reader, for going on it with me.

Notes

PROLOGUE: JOE

3 **rattling, war-weary plane:** Trevor Jermy, "Lt. Joe Kennedy," Norfolk and Suffolk Aviation Museum.
3 **22,000 pounds of British Torpex:** Roger Connor, "Remembering the Death of Lt. Joe Kennedy Jr. and America's First Combat Drones," National Air and Space Museum, August 19, 2014.
3 **world's biggest ever bomb:** Alan Axelrod, *Lost Destiny: Joe Kennedy Jr. and the Doomed WWII Mission to Save London* (St. Martin's, 2015), 178.
4 **realize the dream: a supergun:** Axelrod, *Lost Destiny*, 151.
4 **missiles every minute:** Axelrod, 148.
4 **5,000 men working twenty-four hours a day:** Axelrod, 149.
4 **stopped in the middle of nowhere:** Adam L. Gruen, "Preemptive Defense: Allied Air Power Versus Hitler's V-Weapons, 1943–1945," *Air Force History and Museums Programs*, 1998, 8, 12.
5 **robot arms bolted to the yoke:** Axelrod, *Lost Destiny*, 143.
6 **race in their new sailboat:** Forest Tennant, "John F. Kennedy's Pain Story: From Autoimmune Disease to Centralized Pain," *Practical Pain Management* 12, no. 8 (September 2012).
6 **written up in the newspaper:** Hank Searls, *The Lost Prince: Young Joe, the Forgotten Kennedy* (Ballantine, 1969), 52.
6 **be president one day:** Axelrod, *Lost Destiny*, 55.
6 **parties and sometimes friends:** Axelrod, 56.
7 **teachers called "deficient":** Kate Clifford Larson, *Rosemary: The Hidden Kennedy Daughter* (Houghton Mifflin Harcourt, 2015), 43.
7 **sceney places in Manhattan:** Axelrod, *Lost Destiny*, 54–56.
7 **not a natural pilot:** Axelrod, 53.
7 **twerpy younger brother:** Axelrod, 74.
8 **wave both his hands:** Axelrod, 75–77.
8 **flight has been removed:** Searls, *Lost Prince*, 12.
8 **374 boxes of torpedo explosives:** Jermy, "Lt. Joe Kennedy."
8 **horrible final spasm:** Searls, *Lost Prince*, 252–53.
9 **three-tiered city:** Kathy Myers Krogh and Douglas Myers, *The Professor and the Pilots: Letters Home from Wartime London by a Canadian Psychologist* (FriesenPress, 2018).
9 **control tech is already inside:** Axelrod, *Lost Destiny*, 238, 244.
9 **explosives stacked around him:** Searls, *Lost Prince*, 281–82.

10 **positioned for the takeoff:** "B-17 Pilot Training Manual for the Flying Fortress," AAF Office of Assistant Chief of Air Staff, Training by Headquarters, AAF, Office of Flying Safety, 70.
10 **"Right aileron":** Searls, *Lost Prince,* 247–48.
10 **tech next to him a look:** Axelrod, *Lost Destiny,* 245.
10 **"the rest of my eggs":** Searls, *Lost Prince,* 248.
11 **lifts into the air:** Axelrod, *Lost Destiny,* 247.
11 **chance to help win it:** Axelrod, 235.
11 **president of the United States:** Axelrod, 240.
11 **seen before, even in war:** Steven Russell, "The Kennedy Curse: Tragedy in the Skies over Suffolk," *East Anglian Daily Times,* November 22, 2013.
12 **two words: "Spade Flush":** Searls, *Lost Prince,* 249.
12 **Joe Kennedy in their headsets:** Axelrod, *Lost Destiny,* 194.
12 **plane starts to turn:** Searls, *Lost Prince,* 284.
13 **can see what Joe sees:** Jermy, "Lt. Joe Kennedy."
13 **houses hundreds of yards away:** "Modern Marvels: Smart Bombs," History Channel, September 30, 2003; Searls, *Lost Prince,* 285–86.
13 **scream across the countryside:** Searls, "Lost Prince," 181.
13 **turning the ground to hellfire:** Jack Olsen, *Aphrodite: Desperate Mission* (iBooks, 2004), 117.

BOOK I: HILTON

1.

17 **enemy jets engage:** Richard Hilton, *There Are No Sundays: A Youngster from Oklahoma Finds a Home in High Performance Jet Fighters* (self-published, 2014), 208.
17 **"*that sum' bitch go!*":** Hilton, *There Are No Sundays,* 8–9.
18 **away from terra firma:** Hilton, 13.
18 **panhandle crosses that bridge:** John T. Correll, "The Emergence of Smart Bombs," *Air Force Magazine,* March 2010.

2.

19 **sonar program off the coast:** Author interview with Steve Roemerman; Paul G. Gillespie, *Weapons of Choice: The Development of Precision Guided Munitions* (University of Alabama Press, 2006), 75.
20 **the bridge itself:** Gillespie, *Weapons of Choice,* 77.
20 **converge in his head:** Stephen Coonts and Barrett Tillman, *Dragon's Jaw: An Epic Story of Courage and Tenacity in Vietnam* (Da Capo Press, 2019), 77, 82, 163.
20 **intensity of the sun:** U.S. Army Missile Command, "A Study of the Application of Laser Techniques to Weapon Systems," U.S. Army Missile Command, February 19, 1963, 42.
21 **do much to an enemy:** U.S. Army Missile Command, "Study of Application," 22, 25–26.
21 **"laser-guided stuff?":** "Who Says American Ingenuity Is Dead?," *People,* May 30, 1991.

22 **Nazi regime to America:** Annie Jacobsen, *Operation Paperclip: The Secret Intelligence Program That Brought Nazi Scientists to America* (Little, Brown, 2014); author interview with Steve Roemerman.

22 **was barely aimable:** Alan Axelrod, *Lost Destiny: Joe Kennedy Jr. and the Doomed WWII Mission to Save London* (St. Martin's, 2015), 146.

24 **subsurface magnetic anomaly:** Gary Price, "Texas Instruments," Texas State Historical Association, January 1, 1996.

24 **better binoculars and night vision:** Hans Braakhuis, "The History of Nippon Kogaku, 1600–1949," 2008.

24 **useful to American consumers too:** Author interview with Steve Roemerman.

25 **license it to other companies:** William Brinkman et al., "A History of the Invention of the Transistor and Where It Will Lead Us," *IEEE Journal of Solid-State Circuits* 32, no. 12 (December 1997).

25 **market rather than the current one:** Michael Riordan, "The Lost History of the Transistor," *IEEE Spectrum*, April 30, 2004.

25 **element called germanium:** Riordan, "Lost History."

25 **as abundant as oxygen: silicon:** Dipanjan Sen et al., "Direct Atomistic Simulation of Brittle-to-Ductile Transition in Silicon Single Crystals," MRS Online Proceedings Library 1272, article no. 413 (2010).

25 **still almost nonexistent:** Riordan, "Lost History."

26 **errant test rockets:** "Wernher von Braun," NASA.gov.

27 **destiny around the unlikely marriage:** Author interview with Steve Roemerman.

27 **Midwest to Siberia:** Charles Apple, "The 'Vengeance' Weapon," *Spokesman-Review (Spokane, WA)*, September 2, 2024.

27 **serve as a kind of spokesman:** Jim Denney, "Walt's Vanishing Tomorrowland," *Boardwalk Times*, November 20, 2017.

27 **theme park opening out west:** Bob Gathany, "AL.com Vintage: Disney & Von Braun Inspire America with 'Man in Space,'" *AL.com*, March 9, 2016.

27 **Braun into a household name:** Mike Wright, "The Disney-Von Braun Collaboration and Its Influence on Space Exploration," Southern Humanities Conference, Huntsville, AL, 1993.

28 **shared an uncomfortable connection:** *Time*, February 17, 1958; *Life*, November 18, 1957.

28 **represent an unlimited market:** Author interview with Steve Roemerman.

29 **players in the defense space:** Peter deLeon, "The Laser-Guided Bomb: Case History of a Development," RAND Corporation report prepared for Deputy Chief of Staff, Research and Development (Air Force), June 1974.

29 **making the scale smaller:** deLeon, "The Laser-Guided Bomb."

29 **detachment down at the Florida base:** deLeon.

29 **"chip" into more weapons:** Gillespie, *Weapons of Choice*, 75.

3.

30 **thirty below zero outside:** Author interview with J. D. Franks.

30 **pair of downturned hands:** Author interview with Rick Hilton.

30 **steel coming at him:** Nudelman-Rikhter NR-30, WeaponSystems.net.

31 **not really allowed to fight:** Author interview with Rick Hilton.

31 **politicians have put him in:** Author interview with Rick Hilton.

31 **they can shoot back:** Vago Muradian, "USAF's Operation Bolo Shoots Down Seven North Vietnamese MiF-21s," *Defense & Aerospace Report*, January 2, 2017.
32 **doing just by listening:** Author interview with J. D. Franks; author interview with Rick Hilton.
32 **"Bandits 030 for 15":** Hilton, *There Are No Sundays*, 207–8.
32 **gets ready to fight:** Author interview with Rick Hilton.
32 **shows his distance and heading:** John A. Schell, "The SA-2 and SR-71," April 17, 2020.
33 **big and expensive and unwieldy:** Carlo Kopp, "Engagement and Fire Control Radars (S-Band, X-Band, Ku/K/Ka-Band) Technical Report APA-TR-2009-0102," Air Power Australia, January 2009.
33 **"SEARCH" to a "TRACKING" mode:** Kopp, "Engagement and Fire Control."
33 **right out of the fight:** Author interview with Rick Hilton.
33 **"SAMs! SAMs!":** Hilton, *There Are No Sundays*, 209.
33 **issues a Fire command:** Schell, "SA-2 and SR-71."
34 **reaches the speed of sound:** Coonts and Tillman, *Dragon's Jaw*, 52–53.
34 **dead in the water:** Schell, "SA-2 and SR-71."
34 **out of time to correct:** Hilton, *There Are No Sundays*, 210.
34 **Hilton is standing still:** Bryan Swopes, "22 November 1972," *This Day in Aviation*, November 22, 2024.
34 **spectacular fireworks display:** Hilton, *There Are No Sundays*, 210–12.
35 **space of sensory chaos:** Hilton, 212.
35 **tries not to blink:** Hilton, 210.
36 **recalled all its demons:** Author interview with Rick Hilton.

4.

37 **under a lightning bolt:** Author interview with Rick Hilton.
38 **"I'm not going":** Author interview with Rick Hilton.
38 **"The pitch could be violent":** Author interview with Rick Hilton.
38 **fuel gauge winds down:** Hilton, *There Are No Sundays*, 212.
38 **their most dangerous:** Muradian, "USAF's Operation Bolo."
38 **politicians want to win a war:** Author interview with Rick Hilton.
39 **"got her buck'n'!":** Hilton, There Are No Sundays, 213–14.
39 **"want me to turn?":** Hilton, 214.
39 **"Coming back now":** Hilton, 214–15.
40 **"Breakaway. Breakaway.":** Hilton, 215.
41 **"OK, cut it":** Author interview with Rick Hilton.
42 **landing gear is down:** "Flight Manual USAF Series F-4E Aircraft," McDonnell Aircraft, February 1, 1979.
42 **out of the cockpit:** Hilton, *There Are No Sundays*, 215–16.
42 **"God was with us today?":** Hilton, 217–18.

5.

43 **Satan's Angels moaned:** Author interview with J. D. Franks.
44 **sending tapes back:** Author interview with Rick Hilton.
44 **coming from all directions:** Author interview with Rick Hilton.
45 **overlapping fields of fire:** Coonts and Tillman, *Dragon's Jaw*, 20–22.

45 **wings off the plane:** Author interview with Rick Hilton; Coonts and Tillman, *Dragon's Jaw*, 15.
46 **missile with a television camera:** "Tactical Guided Missiles," *Britannica.com*.
46 **shrapnel from a nearby explosion:** Coonts and Tillman, *Dragon's Jaw*, 33.
46 **Electro-optical bombs lost their targets:** Coonts and Tillman, 213.
46 **happened to the floating mines:** Walter J. Boyne, "Breaking the Dragon's Jaw," *Air & Space Forces Magazine*, August 1, 2011.
46 **radars offline before attacking:** Coonts and Tillman, *Dragon's Jaw*, 223.
47 **get any more tapes:** Author interview with Rick Hilton.

BOOK II: WELDON

1.

51 **curved out toward the sea:** Stephen Coonts and Barrett Tillman, *Dragon's Jaw: An Epic Story of Courage and Tenacity in Vietnam* (Da Capo Press, 2019), 28–33.
51 **existence as Detachment 5:** Peter deLeon, "The Laser-Guided Bomb: Case History of a Development," RAND Corporation report prepared for Deputy Chief of Staff, Research and Development (Air Force), June 1974.
52 **about the military's failure:** deLeon, "The Laser-Guided Bomb."
52 **give him a chance:** Paul G. Gillespie, *Weapons of Choice: The Development of Precision Guided Munitions* (University of Alabama Press, 2006), 77.
52 **no experience in making weapons:** Shelby G. Spires, "Guiding Light," *Smithsonian Air & Space*, May 1999.
52 **asked what lasers were:** Gillespie, *Weapons of Choice*, 106, 109.
52 **had built a chip:** Gillespie, 77–78.
53 **"has a bombing problem":** Gillespie, 77–78.
53 **made Weldon a promise:** Author interview with Paul Gillespie.
53 **"as good as funded":** Gillespie, *Weapons of Choice*, 78.
53 **budget of under $100,000:** Shelby G. Spires, "The TI Paveway Program: The Development," undated.

2.

54 **far too young to die:** "Who Says American Ingenuity Is Dead?," *People*, May 30, 1991.
54 **"you don't want to . . .":** Author interview with Steve Roemerman.
54 **settle his eyes on people:** Author interview with Steve Roemerman.
54 **minute to think:** Author interview with Steve Roemerman.
55 **take shape on paper:** Author interview with Steve Roemerman.
56 **flashlight, at the bridge:** Coonts and Tillman, *Dragon's Jaw*, 208.
56 **bought a six-pack:** Spires, "The TI Paveway Program: The Development."
56 **coast of Rhode Island:** Author interview with Steve Roemerman.
56 **near obsession with data:** "Introducing the Richard H. Johnson Technical Achievement Award," NDIA Conference Proceedings.
57 **something like a divine ideal:** Gillespie, *Weapons of Choice*, 80.
57 **"together and make it work":** "Who Says American Ingenuity Is Dead?"
57 **Weaver's team of ex-Nazis:** deLeon, "The Laser-Guided Bomb."

58 **Even the self-steering weapons:** Author interview with Rick Hilton.
58 **out of a photon detector:** Richard Simons, "The Evolution of Single-Photon Detection," *Laser Focus World*, December 17, 2021.
58 **blast of pressurized gas:** Gillespie, *Weapons of Choice,* 8.
59 **four options available:** R. H. Johnson et al., "Laser-Guided Bomb (Tail Control) Technical Paper," TI Corporate Archives, April 20, 1967, 4–5, 7.
59 ***keep* the eye looking:** Author interview with Steve Roemerman.
60 **"Keep it simple":** Gillespie, *Weapons of Choice,* 80.

3.

61 **working through the weekend:** Gillespie, *Weapons of Choice*, 80.
61 **learned to love data:** "Richard Johnson Obituary," *Dallas Morning News*, July 27, 2008.
61 **The pitot tube, a device:** James L. H. Peck, "How Aircraft Instruments Work," *Popular Science*, March 1944.
61 **badminton birdie so it stabilized itself just by the flow of air:** Gillespie, *Weapons of Choice,* 80- 81.
62 **built and flew in competition:** Peck, "How Aircraft Instruments Work."
62 **he studied the idea:** Peck.
63 **"a Giannini probe!":** Gillespie, *Weapons of Choice,* 80.

4.

64 **officers looked at each other:** Gillespie, *Weapons of Choice*, 81.
64 **Air Force brass was unmoved:** Author interview with Paul Gillespie.
64 **give Weldon $99,000:** Gillespie, *Weapons of Choice,* 81.
65 **If Weldon failed to produce:** Gillespie, 81.
65 **Frankenstein's monster of scrapped:** Gillespie, 83.
65 **all needed wind tunnel time:** Author interview with Steve Roemerman.
65 **sit by his pool:** Gillespie, *Weapons of Choice,* 85–86.
66 **should be stable in flight:** Author interview with Steve Roemerman.
66 **Polaroid was in the middle:** Peter Buse, *The Camera Does the Rest: How Polaroid Changed Photography* (University of Chicago Press, 2016), 31.
66 **Anyone paying even passing:** "Ansel Adams and Polaroid, 1949–1984," Harvard Library.
66 **Rather than inventing a sophisticated:** Author interview with Steve Roemerman.
66 **Soon all the major parts:** Gillespie, *Weapons of Choice,* 87.
67 **gathered at a telemetry pad:** Author interview with Steve Roemerman.

5.

68 **A million things could:** Author interview with Steve Roemerman; author interview with Steve Archibald.
68 **The sensors might not survive:** Gillespie, *Weapons of Choice*, 87–88.
68 **There was an awkward silent:** Author interview with Steve Roemerman.
69 **When they found the bomb:** deLeon, "The Laser-Guided Bomb."
69 **they found the bomb plugged:** deLeon.
69 **The fins had moved:** Author interview with Steve Roemerman.

6.

70 **SAMs and 85's:** Author interview with Richard Hilton.
70 **He's straining to get:** Philip D. Chinnery, *Full Throttle* (St. Martin's Press, 1988).
70 **There are sixteen planes:** Coonts and Tillman, *Dragon's Jaw,* 237–38.
71 **not just clouds; it's haze:** Author interview with Rick Hilton.
71 **One of the planes:** Coonts and Tillman, *Dragon's Jaw,* 238.
72 **"Roger, making a port":** Coonts and Tillman, 238.
72 **57mm fast-firing antiaircraft:** Coonts and Tillman, 238.

7.

73 **Weldon looked over:** Author interview with Steve Roemerman.
73 **He considered the design:** Gillespie, *Weapons of Choice*, 81.
74 **Again they found the bomb:** deLeon, "The Laser-Guided Bomb."
74 **an old two-and-a-half-ton Army truck:** Gillespie, *Weapons of Choice,* 88.
74 **landed twelve feet away:** deLeon, "The Laser-Guided Bomb."
74 **landed within ten feet:** deLeon.
75 **This larger bomb test:** Gillespie, *Weapons of Choice,* 81.
75 **Before this final test:** Gillespie, 89.
75 **The little pedestrian bridge:** deLeon, "The Laser-Guided Bomb."

8.

76 **Back at the Pentagon:** Gillespie, *Weapons of Choice*, 107–8.
76 **The colonel didn't have:** Gillespie, 107–8.
76 **When Weldon walked into:** Gillespie, 106–8.
77 **LeMay was famous for:** Seth Paridon, "Hellfire on Earth: Operation MEETINGHOUSE," WWII: The National WWII Museum, March 8, 2020.
77 **planes to drop "LeMay leaflets":** "Warning Leaflets," Atomic Heritage Foundation, 2022.
77 **Now, Weldon was presenting:** Gillespie, *Weapons of Choice*, 107.
77 **By using laser designators:** Gillespie, 88.
78 **LeMay puffed; he seemed:** Gillespie, 107.
78 **The Aeronautical Systems Division:** deLeon, "The Laser-Guided Bomb."
78 **It was a final adjustment:** Gillespie, *Weapons of Choice,* 110.
79 **More ideas emerged:** Gillespie, 130.
79 **The first reports coming:** Barry Miller, "Avionics: Tactical Laser Use Advancing: Devices Gain Growing Weapons Role," *Aviation Week & Space Technology*, January 19, 1970.
79 **Memos back from the field:** Spires, "Guiding Light"; Miller, "Avionics: Tactical Laser Use Advancing: Devices Gain Growing Weapons Role."
79 **President Johnson was in a predicament:** John T. Correll, "The Emergence of Smart Bombs," *Air Force Magazine*, March 2010.
79 **The military had to compensate:** Gillespie, *Weapons of Choice,* 112.
79 **team could make 1,000:** deLeon, "Laser-Guided Bomb."
79 **2,000 a month:** Gillespie, *Weapons of Choice*, 115–16.
80 **The South Vietnamese military:** Michael Casey et al., *Flags into Battle* (Boston Publishing Company, 1987), 182.

9.

81 **That mission failed:** Author interview with Rick Hilton.
81 **They'd accomplished nothing:** Author interview with Rick Hilton.
81 **It took two weeks:** A. J. C. Lavalle, *The Tale of Two Bridges and the Battle for the Skies of North Vietnam* (Office of Air Force History, 1985).
82 **The four formations bob:** Author interview with Rick Hilton.
82 **He's lost the bridge:** Author interview with Rick Hilton.
82 **"If you have the target":** Author interview with Rick Hilton.
82 **Hilton watches the other:** Coonts and Tillman, *Dragon's Jaw*, 238.
82 **The plan is for Hilton:** Melvin F. Porter, *Linebacker: Overview of the First 120 Days* (Headquarters, Pacific Air Forces, September 1972).
83 **"OK, Goatee," he says:** Author interview with Rick Hilton.
83 **A beat of dead air:** Author interview with Rick Hilton.
84 **The plane hums:** Author interview with Rick Hilton.
84 **Thirty-seven thousand pounds:** Author interview with Rick Hilton.
84 **As Hilton pulls up, the force:** Author interview J. D. Franks; author interview with Rick Hilton.
85 **No sound at all:** Author interview with Bill Wideman; author interview with Rick Hilton.
86 **He's recording a new:** Coonts and Tillman, *Dragon's Jaw,* 242.

BOOK III: WIEGAND

1.

89 **Crammed with families living:** John O. Koehler, *Stasi: The Untold Story of the East German Secret Police* (Westview Books, 1999), 329.
90 **Something sinister was happening:** *Frontline*, "My Brothers Bomber," PBS, 2015.
91 **As Wiegand watched:** Koehler, *Stasi*, 327.

2.

92 **Air Force planners calculated:** Melvin F. Porter, *Linebacker: Overview of the First 120 Days* (Headquarters, Pacific Air Forces, September 1972).
93 **Against hard targets, Paveway proved:** John T. Correll, "The Emergence of Smart Bombs," *Air Force Magazine*, March 2010.
93 **Texas Instruments had delivered 100,000 Paveways:** Paul G. Gillespie, *Weapons of Choice: The Development of Precision Guided Munitions* (University of Alabama Press, 2006), 129.
93 **Texas Instruments was ramping up:** Author interview with Steve Roemerman.
93 **Working the Paveway assembly:** Author interview with Steve Roemerman.
94 **There were studies suggesting:** Liudmila Liutsko et al., "Fine Motor Precision Tasks: Sex Differences in Performance with and Without Visual Guidance Across Different Age Groups," *Behaviorial Sciences* 10, no. 1 (January 2020).
94 **Along with a new influx:** Fiona M. Wilson, *Organizational Behavior and Gender: A Critical Introduction* (Oxford University Press, 1999).
94 **Competitors tried to break:** Gillespie, *Weapons of Choice,* 130.

95 **Paveway had to remain simple:** Author interview with Steve Roemerman.
95 **"This needs to be about":** Author interview with Steve Roemerman.

3.

96 **Libya seemed to be lurking:** Terry Atlas and James O'Shea, "Libyan Diplomats Stalk Americans," *Chicago Tribune*, April 13, 1986.
96 **In '73, Libya was linked:** "Where Two U.S. Envoys Died in Search of Peace," *U.S. News & World Report*, June 28, 1976.
96 **in '75, when militants stormed:** Clyde H. Farnsworth, "Terrorists Raid OPEC Oil Parley in Vienna, Kill 3," *New York Times*, December 22, 1975.
96 **And now the CIA:** Peter Kihss, "Qaddafi Denies a Murder Plot; U.S. Is Insistent," *New York Times*, December 7, 1981; Gregory L. Trebon, "Libyan State Sponsored Terrorism: What Did Operation El Dorado Canyon Accomplish?," Air Command and Staff College, June 1986.
96 **Thanks to Libya, Reagan:** Edward C. Keefer, *Caspar Weinberger and the U.S. Military Buildup, 1981–1985* (Historical Office of the Secretary of State, 2023).
96 **Reagan believed he was facing:** National Security Council Meeting minutes, December 8, 1981.
97 **"With all due respect":** "Summary of National Security Planning Group (NSPG) Meeting on Combatting Terrorism," March 2, 1984.
98 **"The JCS have the same":** "Summary of National Security Planning Group."

4.

99 **Berlin was still the city:** Reinhard Joksch, *Berlin, Berlin* (TV documentary series), DOKfilm Fernsehproduktion, August 2015.
99 **There were 200,000:** Tobias Wunschik, "Political Prisoners in the German Democratic Republic," *Communist Crimes*, May 27, 2020.
99 **Recording devices were hidden:** Christian Gierke, *Stasi: East Germany's Secret Police* (documentary), Film Europa, 2007.
99 **Beneath Wiegand's office, windowless trucks:** Emine Seda Kayim, "Selective Abolitionism: Germany's Socialist Prisons and 'Coming-to-Terms' with One Side of History," *Platform*, August 12, 2024.
99 **Scent samples from citizens:** Gierke, *Stasi: East Germany's Secret Police.*
99 **Beyond the Stasi's 90,000:** Kate Connolly, "The Machine That Is Putting Together the Stasi's 600m-Piece Spy Jigsaw," *Guardian*, May 10, 2007.
100 **By proportion of the population:** Gierke, *Stasi: East Germany's Secret Police.*
100 **The giant white domes:** Sam Shead, "Teufelsberg: The Abandoned US Spy Station in Berlin That's Been Turned into a Graffiti Artists' Playground," *Business Insider*, June 16, 2016.
101 **The Libyans were planning something:** Koehler, *Stasi*, 332.
101 **The new mole, code-named "Mustafa":** Koehler, 336–37.
102 **He got back to HQ:** Koehler, 146.
102 **Wiegand's friend Wolfgang Stuchly:** Koehler, 329.
102 **When Stuchly arrived:** Koehler, 337.
103 **Alba was getting close to Musbah:** Koehler, 329, 337.
103 **Wiegand was not just:** Koehler, 330.

103 **"We are not murderers!":** Koehler, 329.
103 **They warned him to leave:** Koehler, 331–33.
104 **his best confidential informants:** Koehler, 333.

5.

105 **Two weeks after Reagan:** Michele Neubert and F. Brinley Bruton, "Yvonne Fletcher Case: Libyan Arrested in Shooting at Embassy," NBC News, November 19, 2015.
106 **At approximately 9:10:** John Tagliabue, "Airport Terrorists Kill 13 and Wound 113 at Israeli Counters in Rome and Vienna," *New York Times*, December 28, 1985; Loren Jenkins, "17 Die in Attacks on Rome, Vienna Airports," *Washington Post*, December 27, 1985.
106 **"I think we're all":** Ronald Reagan, *The Reagan Diaries* (HarperCollins, 2007), 381.
106 **They already had proof:** Judy G. Endicott, "Raid on Libya: Operation Eldorado Canyon," in *Short of War: Major USAF Contingency Operations, 1947–1997*, ed. A. Timothy Warnock (Air University Press, 2000), 149–50.
107 **Even more reluctantly this time:** Reagan, *Reagan Diaries,* 381.

6.

108 **New threat intel landed:** Koehler, *Stasi*, 335.
108 **Intercepted cable traffic:** R. C. Haerr, "The Gulf of Sidra," *San Diego Law Review*, 1987, 753–54.
108 **Wiegand read surveillance:** Ruling of the Berlin Court, conviction of the perpetrators of the La Belle Discotheque Bombing, April 5, 1986, Berlin; Koehler, *Stasi,* 334.
108 **And Wiegand's one reluctant:** Koehler, 335.
108 **A plot to use grenades:** Koehler, 334–35.
109 **The U.S. Army's hospital:** Ruling of the Berlin Court.
109 **Alba reported having himself:** Ruling of the Berlin Court.
109 **An action was to be taken:** Koehler, *Stasi,* 335.
109 **Wiegand grew frantic:** Koehler, 335–36.
110 **"They could turn against *us*":** Koehler, 334–35.
110 **He'd dispatch everyone:** Koehler, 336.
110 **The day before the expected:** Koehler, 336–37.
110 **Two employees from the Libyan:** Ruling of the Berlin Court.
110 **Volkswagen turned in to the Kreuzberg:** Ruling of the Berlin Court.
111 **He said other intelligence services:** Koehler, *Stasi,* 336–37.
112 **There was finally something:** Koehler, 338.
112 **Musbah Eter was carrying:** Ruling of the Berlin Court.

7.

113 **At twenty-six years old, Musbah:** Veronica Nmoma, "Power and Force: Libya's Relations with the United States," *Journal of Third World Studies* 26, no. 2 (2009): 137–59.
113 **By the time Eter was:** Nmoma, "Power and Force," 137–59.
114 **Eter was a polymath:** Ruling of the Berlin Court, conviction of the perpetrators of the La Belle Discotheque Bombing, April 5, 1986, Berlin.
114 **He'd fallen ill:** Ruling of the Berlin Court.
114 **Now, two years later:** Ruling of the Berlin Court.

114 **When Eter snuck the detonator:** Harris et al. v. Socialist People's Libyan Arab Jamahiriya et al., United States District Court for the District of Columbia, April 21, 2006.
115 **Ali Chanaa was also:** Koehler, *Stasi*, 334.

8.

116 **American pilots training at Nellis:** Author interview with Steve Roemerman.
116 **Nellis Air Force base:** Author interview with Steve Roemerman.
117 **TI decided an employee:** Author interview with Steve Roemerman.
117 **Weldon considered the reports:** Author interview with Steve Roemerman.
117 **A plane equipped to avoid:** Correll, "The Emergence of Smart Bombs."
118 **If Paveway was going:** Gillespie, *Weapons of Choice*, 130–31.
119 **It had taken two decades:** Gillespie, 91.
119 **The newer integrated circuit microchip:** Author interview with Steve Roemerman.
120 **With a maturing body:** Author interview with Steve Roemerman.

9.

121 **Soon after Wiegand and Stuchly:** Ruling of the Berlin Court, conviction of the perpetrators of the La Belle Discotheque Bombing, April 5, 1986, Berlin.
121 **Wiegand called Stuchly back:** John O. Koehler, *Stasi* (Basic Books, 1999) Koehler, *Stasi*, 333.
122 **One by one, he called:** Koehler, 338–39.

10.

123 **1,500 grams of plastic:** Ruling of the Berlin Court, conviction of the perpetrators of the La Belle Discotheque Bombing, April 5, 1986, Berlin.
123 **Eter turned to Verena:** Ruling of the Berlin Court.
124 **Wouldn't two women be less:** Karsten Gaede, "Case Law: HRRS 2004 No. 727," June 24, 2004.
124 **He slipped the bomb:** Ruling of the Berlin Court.
124 **Eter crossed back into:** Mary Williams Walsh, "Germany Finally to Try '86 Disco Bombing Case," *Los Angeles Times*, November 18, 1997.

11.

125 **"tripoli will be happy":** David C. Wills, *The First War on Terrorism: Counter-Terrorism Policy During the Reagan Administration* (Rowman & Littlefield, 2003), 196–97.
125 **The military sent an urgent:** Joseph Stanik, *El Dorado Canyon: Reagan's Undeclared War with Qaddafi* (Naval Institute Press, 2002); Stewart Powell and Robert A. Kittle, "Can Reagan Make Qadhafi Cry Uncle?," *U.S. News & World Report*, April 21, 1986.
125 **Decoded and translated, it read:** Koehler, *Stasi*, 326.

12.

126 **Just before midnight:** Ruling of the Berlin Court, conviction of the perpetrators of the La Belle Discotheque Bombing, April 5,1986; Peter Finn, "4 Guilty in '86 Disco Blast in Germany," *Washington Post*, November 4, 2001.

126 **They approached a six-story:** John Tagliabue, "2 Killed, 155 Hurt in Bomb Explosion at Club in Berlin," *New York Times*, April 6, 1986.
126 **The purer American targets:** Ruling of the Berlin Court.
126 **He'd decided La Belle:** Ruling of the Berlin Court.
126 **People began to file:** Koehler, *Stasi*, 356.
127 **Even from a distance:** Walsh, "Germany Finally to Try '86 Disco Bombing Case."
127 **Hundreds of eardrums burst:** Katja Bahadori, "April 5, 1986: Attack on La Belle Nightclub" (Contemporary Witnesses, Politics, West), The Berlin Wall: A Multimedia History.

13.

128 **"I believe this *is*":** Caspar Weinberger, *Fighting for Peace: Seven Critical Years at the Pentagon* (Michael Joseph, 1990), 188.
128 **made it back to Washington:** Weinberger, *Fighting for Peace,* 132–33.
128 **In the Situation Room:** Author interview with Steve Roemerman.
128 **The generals had exactly:** Endicott, "Raid on Libya," 145–55.
128 **pilots had practiced extensively:** Author interview with Steve Roemerman.
129 **now stationed in Lakenheath, England:** Endicott, "Raid on Libya."
129 **President Reagan approved:** "Congressmen Back Reagan's Move on Libya," *Los Angeles Times*, April 16, 1986.
129 **The operation, code-named:** Walter J. Boyne, "El Dorado Canyon," *Air Force Magazine*, March 1999.

14.

130 **Without cooperation from:** Boyne, "El Dorado Canyon."
131 **six Aardvarks buzzed the Tripoli:** Endicott, "Raid on Libya," 145–55.
131 **The raid was frenetic:** Endicott.

15.

132 **A rumor passed through:** Seymour M. Hersh, "TARGET QADDAFI," *New York Times Magazine*, February 22, 1987.
132 **Many Americans listened to it:** Robert McFadden, "In the U.S., Audiences Listen In on the Attacks," *New York Times*, April 15, 1986.
133 **The loss of one plane:** Endicott, "Raid on Libya," 145–55; Boyne, "El Dorado Canyon."
133 **child named Hana:** Patrick Müller, "Is Gaddafi's Daughter, Believed Killed by a U.S. Air Strike, Alive and Well?," *Die Welt*, August 12, 2011.
133 **Baby Hana became:** "My Brother's Bomber" (documentary), *Frontline*, PBS, 3 episodes, aired September 29, October 6, and October 13, 2015.
134 **"Reagan," the papers said, "was magnificent":** Howard Rosenberg, "Ronald Reagan: America's Anchorman: 'Even When He Makes a Terrible Joke or Speaks a Sentence That Has No Beginning and End, He Seems Like a "Nice" Person,'" *Los Angeles Times*, April 16, 1986.
134 **"In a credibility contest":** Rosenberg, "Ronald Reagan."
134 **"Failure to act after":** Keith Love, "Cranston, GOP Senate Candidates Support President on Libya Attack," *Los Angeles Times*, April 16, 1986.
134 **The success of the mission:** Bill Clinton: *My Life* (Vintage, 2005), 717.

135 **nightclub bombing killed:** Steven Erlanger, "4 Guilty in Fatal 1986 Berlin Disco Bombing Linked to Libya," *New York Times,* November 14, 2001.
135 **Border guards searched the box:** Ruling of the Berlin Court, conviction of the perpetrators of the La Belle Discotheque Bombing, April 5, 1986.
135 **His applications for travel:** Ruling of the Berlin Court.
136 **walked into the German embassy:** Ruling of the Berlin Court; Harris et al. v. Socialist People's Libyan Arab Jamahiriya et al., United States District Court for the District of Columbia, April 21, 2006.

16.

138 **The day after the raid:** Author interview with Steve Roemerman.
139 **He wasn't happy, though:** Author interview with Rick Hilton.
139 **Karma 51 had crashed:** Ruth Marshall, "View from the Bull's-Eye," *Newsweek,* April 28, 1986.

BOOK IV: KATHY

1.

143 ***Hey, Missiles*, they say:** Kathy Kelly, *Other Lands Have Dreams: From Baghdad to Pekin Prison* (CounterPunch, 2005), 93.
143 **protest movement she'd practiced:** Kelly, *Other Lands Have Dreams,* 21.
144 **Down there in the dark:** Mead & Hunt, "The Missile Plains: Frontline of America's Cold War," United States Department of the Interior, National Park Service, Midwest Regional Office, 2003, 31.
144 **That this weapon, the Minuteman:** Author interview with Kathy Kelly.
145 **She sits on the concrete:** Kelly, *Other Lands Have Dreams,* 22.
145 **She catches bits of transmissions:** Kelly, 22.
146 **The thousand silos marching:** "Minuteman Missile National Historic Site: Protecting a Legacy of the Cold War (Teaching with Historic Places)," National Park Service.
147 **The soldier breathes behind her:** Kelly, *Other Lands Have Dreams,* 22.
147 **She reported to county lockup:** Kelly, 92.
147 **Her eye itched, her arm throbbed:** Kelly, 92.
147 **She fought tears:** Author interview with Kathy Kelly.
148 **In jail, on the phone:** Author interview with Kathy Kelly.
149 **"I tried my hardest":** Kelly, *Other Lands Have Dreams,* 92–93.
150 **Hiding sobs behind giant:** Kelly, 112.
150 **She'd carry a guilt:** Kelly, 92.
151 **"Soldiers don't really have":** Kelly, 14–15.

2.

153 **Right around the time:** William M. Arkin, "Baghdad: The Urban Sanctuary in Desert Storm?," *Airpower Journal*, Spring 1997; Rick Atkinson, "Architects of the Air War," *Washington Post*, October 2, 1993.
153 **This time he'd use Checkmate:** Eliot A. Cohen et al., "Gulf War Air Power Survey, vol. 1—Planning and Command and Control," Library of Congress, 1993.

153 **Now, Colonel John Warden:** John Andreas Olsen, "Warden Revisited: The Pursuit of Victory Through Air Power," *Air Power History* (Winter 2017), 41.
154 **The way Warden saw it:** Rick Atkinson, *Crusade: The Untold Story of the Persian Gulf War* (Houghton Mifflin, 1993), 272–74.
154 **Like the Operation El Dorado:** Cohen et al., "Gulf War Air Power Survey."
155 **Once Colonel Warden's Checkmate:** William M. Arkin, "Masterminding an Air War," *Washington Post*, 1998.
155 **the company's first foreign factory:** Author interview with Steve Roemerman.
155 **vaulted by his offspring's success:** Author interview with Steve Archibald.
155 **He still thought Paveway's:** Author interview with Steve Roemerman.
156 **New enemy methods:** Raytheon (Texas Instruments) Paveway III History and General Description.
156 **Weldon challenged his team:** Author interview with Steve Roemerman.
157 **Paveway I and II flew:** John T. Correll, "The Emergence of Smart Bombs," *Air Force Magazine*, March 1, 2010.
157 **the Air Force turned out:** Author interview with Steve Roemerman.
157 **"Paveway is a pretty nice":** Author interview with Steve Roemerman.
158 **But Weldon had earned:** Author interview with Steve Roemerman.
158 **He didn't. And almost:** Author interview with Steve Roemerman.
159 **But when the auditors:** Author interview with Steve Roemerman.
159 **The Air Force had been running:** Author interview with Steve Roemerman.
159 **contracted with the secret Skunk Works:** Author interview with Steve Roemerman.
159 **Pilots flying a true stealth:** Author interview with Steve Roemerman.
160 **Without realizing it, Weldon's team:** Author interview with Steve Roemerman.

3.

161 **Colonel John Warden began:** Cohen et al., "Gulf War Air Power Survey."
161 **He sent a contingent of Checkmate:** Atkinson, "Architects of the Air War"; Cohen et al., "Gulf War Air Power Survey."
161 **two cells established a protocol:** Atkinson, "Architects of the Air War."
161 **began adding potential target:** Atkinson, 275.
162 **Finding backup facilities:** Atkinson, 275.
162 **Black Hole staffers began tracking down:** Arkin, "Baghdad: The Urban Sanctuary in Desert Storm?"; Atkinson, "Architects of the Air War."
162 **Among the most noteworthy:** Atkinson, "Architects of the Air War," 275.

4.

163 **Kathy Kelly scrambled:** Author interview with Kathy Kelly; Arthur E. Rowse, "How to Build Support for a War," *Columbia Journalism Review*, September–October 1992.
163 **Among those viewers:** Author interview with Kathy Kelly.
164 **Kathy step touched at a debate:** Author interview with Kathy Kelly.
164 **Saddam's army had pillaged:** Glenn Frankel, "The Other Gulf War—over Money," *Washington Post*, August 19, 1990.
164 **PSAs began to fill:** Atkinson, *Crusade*, 27.
165 **Behind a sign that read:** Robert D. McFadden, "Iraqis Assail U.S. as Rescue Goes On," *New York Times*, February 15, 1991.

165 **To children, Public Shelter No. 25:** Atkinson, *Crusade,* 275; "The Bombing of Iraqi Cities: Middle East Watch Condemns Bombing Without Warning of Air Raid Shelter in Baghdad's Al Ameriyya District on February 13," *Middle East Watch*, March 9, 1991.

165 **As parents began:** Alfonso Rojo, "Bodies Shrunk by Heat of Fire," *Guardian*, February 14, 1991.

5.

166 **Two hundred miles southeast:** Author interview with Kathy Kelly.

166 **Still, they prepared:** Kathy Kelly, "The Children of Iraq," *Link*, January–March 1997.

167 **They were seventy-three:** Kelly, "The Children of Iraq."

6.

169 **Just before 3 a.m.:** James P. Coyne, "A Strike by Stealth," *Air & Space Forces*, March 1, 1992.

169 **"Don't grandstand this one":** Peter Arnett, *Live from the Battlefield: 35 Years in the World's War Zones* (Simon & Schuster, 1994), 364.

170 **Out in the desert:** Author interview with Kathy Kelly.

171 **But now, the signs of destruction:** Kelly, "The Children of Iraq."

172 **But she felt OK:** Author interview with Kathy Kelly.

172 **When the buses downshifted:** Author interview with Kathy Kelly.

172 **In Baghdad, inside Public Shelter:** Author interview with Mohammed Al Qaisi.

172 **a time of separated families:** Kelly, "The Children of Iraq."

173 **Regardless of what happened:** Atkinson, *Crusade*, 275; Rojo, "Bodies Shrunk by Heat of Fire."

173 **regime was clearly losing control:** Kevin M. Woods, "Iraqi Perspectives Project Phase II Um Al-Ma'arik (The Mother of All Battles): Operational and Strategic Insights," Institute for Defense Analyses, May 2008.

7.

174 **Once Black Hole staffers:** Atkinson, *Crusade*, 275.

175 ***Everything* in Baghdad:** Atkinson, 276.

175 **"Activated, recently camouflaged":** Atkinson, 276–77.

8.

176 **At an air base wedged:** Coyne, "A Strike by Stealth"; Richard G. Davis, "On Target: Organizing and Executing the Strategic Air Campaign Against Iraq," United States Air Force History and Museums Program, 2002.

176 **They secured the bombs:** Author interview with Steve Archibald.

176 **The bombers' downward-facing:** Douglas G. Adler, "The F-117 Is the Stealthy Fighter You Can Now See for Yourself," *Air Force Times*, October 30, 2023.

177 **Heat 34 and Rain 37 arrived:** Coyne, "A Strike by Stealth."

178 **He pressed the CONSENT button:** Coyne.

178 **The pilots made a programmed:** Maura Stephens, "Letter from Baghdad," *Salon*, March 20, 2003.

9.

179 **A breeze eased against Kathy's:** Author interview with Kathy Kelly.
180 **Kathy called the American ambassador:** Author interview with Kathy Kelly.
180 **Embedding with the convoys:** Kelly, *Other Lands Have Dreams*, 39–41.
181 **It turned out that Nayirah:** Rowse, "How to Build Support for War," 28–29.
182 **first thing that Kathy noticed:** Kelly, "The Children of Iraq."
182 **She followed the banners:** Maggie O'Kane, "The Most Pitiful Sight I Have Ever Seen," *Guardian*, February 14, 2003.
182 **Parents had delivered their children here:** Rojo, "Bodies Shrunk by Heat of Fire."
183 **flowers were probably plastic:** John Simpson, *The Wars Against Saddam: Taking the Hard Road to Baghdad* (Macmillan, 2003).
183 **As she walked back:** Kelly, "The Children of Iraq."
183 **He'd decided he'd reached:** "Who Says American Ingenuity Is Dead?," *People*, May 30, 1991.
184 **But while other Americans:** Hussein Amin, "Watching the War in the Arab World," Transnational Broadcasting Studies, no. 10 (Spring/Summer 2003); "Television—the Persian Gulf War," *Encyclopedia of the New American Nation*.
184 **He'd always known he was:** "Who Says American Ingenuity Is Dead?"
184 **Now that one-night raid:** "Who Says American Ingenuity Is Dead?"
184 **Weldon tried to distract:** "Who Says American Ingenuity Is Dead?"

BOOK V: RADICA

1.

187 **before the Soviet Union collapsed:** Author interview with Steve Roemerman.
187 **A bomber invisible to radar:** John T. Correll, "The Emergence of Smart Bombs," *Air Force Magazine*, March 1, 2010.
187 **If anything, it probably saw worse:** Justin Hemann, "Case Study Report: The Paveway Program Transformation," unpublished working paper, Lean Aerospace Initiative, Massachusetts Institute of Technology, February 28, 2005.
188 **To TI execs, it wasn't:** Mark C. Cleary, "The Cape: Military Space Operations, 1971–1992," *Military Space Program Guide*, January 1994.
189 **Pat Haggerty figured:** Author interview with Steve Roemerman.
189 **Together, the arguments worked:** Author interview with Steve Roemerman.

2.

191 **It's December, just before:** Bradford Lyttle, "Solidarity for Peace in Sarajevo," *Peace Activist*, 1992.
191 **She's on her way to Sarajevo:** Author interview with Kathy Kelly.
192 **nearly five hundred:** Lyttle, "Solidarity for Peace in Sarajevo."
193 **The team endures the storm:** Lyttle.
194 **They pass homemade shields:** Lyttle.
195 **She decides to invite:** Author interview with Kathy Kelly.
196 **state-of-the-art weapons facility:** "UK Link to Serbian Poison Gas," *Guardian*, April 24, 1999.

3.

197 **Everything needed to mount:** "Former Yugoslavia Chemical Overview," NTI (Nuclear Threat Initiative), June 10, 2014.
197 **The Mostar facility produced:** "Former Yugoslavia Chemical Overview."
198 **But here, with a daughter:** Author interview with Radica Nikolic.

4.

199 **Among the few people ambitious:** Bill Clinton, *My Life* (Alfred A. Knopf, 2004), 717.
199 **Next to Bush, Clinton:** Ian Traynor and Maggie O'Kane, "The Siege of Sarajevo—Archive, 1993," *Guardian*, July 18, 2018.
200 **Only when Serb forces:** Michael O. Beale, *Bombs over Bosnia: The Role of Airpower in Bosnia-Herzegovina* (Air University Press, 1997), 33.
201 **By 1995, Clinton decided:** Ivo H. Daalder, "Decision to Intervene: How the War in Bosnia Ended," *Brookings*, December 1, 1998.
201 **Clinton called in Tony Lake:** Daalder, "Decision to Intervene."
201 **He launched a series:** Monica Hanson Green, "Srebrenica Genocide Denial Report 2020," Srebrenica Memorial, May 2020, 14.
202 **NATO leaders fell into place:** Anthony M. Schinella, *Bombs Without Boots: The Limits of Airpower* (Brookings Institution Press, 2019), 37.

5.

204 **Bids began to come in:** Evan Ramstad, "TI-Raytheon Agreement Involves Shared Campus," *Wall Street Journal*, January 29, 1997.
205 **They'd also just acquired:** "Raytheon in Tucson," *Arizona Daily Star*, December 4, 2011.
205 **Raytheon offered relocation:** Author interview with Steve Roemerman.

6.

206 **Clinton's air campaign:** Charles King, "The Bad Guy" (book review of *Milosevic: Portrait of a Tyrant* by Dusko Doder and Louise Branson), *New York Times*, November 21, 1999.
206 **Despite Kosovo's place in Serb:** Harvey W. Kushner, *Encyclopedia of Terrorism* (Sage Publications, 2002), 20.
207 **"If we have to use force":** Secretary of State Madeleine K. Albright, interview on *Today* with Matt Lauer, NBC-TV, February 19, 1998.

7.

209 **They had to assemble Paveway:** Author interview with Steve Roemerman.
209 **Even machinists ran afoul:** Hemann, "Case Study Report: The Paveway Program Transformation."
209 **Paveway units started coming:** Author interview with Steve Archibald.
210 **Weldon's protégé asked:** Author interview with Steve Roemerman.
210 **Weldon's protégé looked:** Author interview with Steve Roemerman.

8.

212 **On March 24, 1999, Clinton:** Statement by the President to the Nation, Clinton White House Archives, March 24, 1999.

9.

213 **By four years old, Kiki:** Author interview with Ljubica "Kiki" Nikolic.

10.

215 **It was a warm, mid-spring:** Author interview with Radica Nikolic.

11.

219 **"a victory for a safer world":** Statement by the President to the Nation, Clinton White House Archives, June 11, 1999.

12.

220 **Radica could see, even:** Author interview with Radica Nikolic.
220 **Around her, people began:** Serhiy Kurykin, "Environmental Impact of the War in Yugoslavia on South-East Europe," Report of the Committee on the Environment, Regional Planning and Local Authorities, January 10, 2001.
221 **But Kiki hated the move:** Author interview with Ljubica "Kiki" Nikolic.
221 **When Radica gave birth:** Author interview with Radica Nikolic.
222 **Kathy Kelly sits at her father's:** Author interview with Kathy Kelly.
222 **Old friends come by:** Kathy Kelly, *Other Lands Have Dreams: From Baghdad to Pekin Prison* (CounterPunch, 2005), 24.
223 **"I just wonder what would":** Author interview with Kathy Kelly.
223 ***How do you even pay*:** Kelly, *Other Lands Have Dreams,* 24.
224 **Campaigning on the notion:** Dan Balz, "Bush Favors Internationalism," *Washington Post*, November 19, 1999.

BOOK VI: LUIS

1.

227 **He'd spent the bulk:** Author interview with Luis Rueda.
228 **hell was "Persian Gulf Affairs":** Author interview with Luis Rueda.
229 **They began publishing journal articles:** Letter to the Honorable William J. Clinton from Project for the New American Century, January 26, 1998.
229 **"From the Gulf War to Operation":** *Rebuilding America's Defenses: Strategy, Forces and Resources for a New Century*, a Report of the Project for the New American Century, September 2000.
229 **Rueda knew the CIA:** Author interview with Luis Rueda.
229 **When Rueda started:** Author interview with Luis Rueda.

230 **The last time the CIA had tried:** Scott Ritter, "The Coup That Wasn't," *Guardian*, September 27, 2005.
230 **He had every single CIA:** James Risen, "FBI Probed Alleged CIA Plot to Kill Hussein," *Los Angeles Times*, February 15, 1998.
230 **Many of the officers involved:** Author interview with Luis Rueda.
230 **He looked everywhere:** Author interview with Luis Rueda.
230 **Saddam had body doubles:** Tom Zeller, "Will the Real Saddam Hussein Please Step Down," *New York Times*, October 6, 2002.
230 **He sometimes moved:** Donald Rumsfeld, *Known and Unknown* (Penguin, 2011), 460.
230 **He made the rounds among:** Author interview with Luis Rueda.
231 **Rueda was a month:** Author interview with Luis Rueda.
232 **At Langley, the Counterterrorism:** George Tenet, *At the Center of the Storm* (HarperCollins, 2007), 306–7.
232 **"Iraq? Why the hell":** Author interview with Luis Rueda.
232 **McLaughlin wanted to help:** Author interview with Luis Rueda.

2.

234 **Forty years before, a four-year-old:** Author interview with Luis Rueda.
234 **Everything was strange:** Author interview with Luis Rueda.
235 **The family finally reunited:** Author interview with Luis Rueda.
236 **A CIA-backed brigade:** "The Bay of Pigs: Lessons Learned," JFK Library.
236 **Luis's father hadn't:** Author interview with Luis Rueda.
237 **Luis watched his father:** Author interview with Luis Rueda.
237 **reason to be nervous:** Author interview with Luis Rueda.
238 **Rueda's training in clandestine:** Author interview with Luis Rueda.
238 **Now, five tours in:** Author interview with Luis Rueda.

3.

240 **Rueda decided he'd start:** Author interview with Luis Rueda.
240 **He got the money:** Author interview with Luis Rueda.
240 **He'd need a crack team:** Author interview with Luis Rueda.

4.

242 **By the early fall of 2001:** Author interview with Steve Archibald.
242 **The Paveway assembly line in Arizona:** Justin Hemann, "Case Study Report: The Paveway Program Transformation," unpublished working paper, Lean Aerospace Initiative, Massachusetts Institute of Technology, February 28, 2005.
242 **One of the few blemishes:** Sebastian Roblin, "Stealth Shootout: In 1999, a U.S. Air Force F-117 Was Shot Down. Here's How It Went Down," *National Interest*, December 3, 2018.
243 **Stealth technology being defeated:** Darrel Whitcomb, "The Night They Saved Vega 31," *Air Force*, December 2006.
243 **Two potential solutions:** "Raytheon Demonstrates Enhanced Paveway II Data Link Capability," RTX, July 18, 2006.

243 **Boeing had already introduced:** "GBU-39B Small Diameter Bomb Weapon System," Air Force.

243 **Before Raytheon acquired Paveway:** Eric Long, "TI-4100 NAVSTAR Navigator GPS Receiver," National Air and Space Museum, Smithsonian Institution.

243 **TI's GPS technology:** Author interview with Steve Roemerman.

244 **To realize that dream:** Author interview with Steve Roemerman.

245 **The plant in Arizona:** Hemann, "Case Study Report: The Paveway Program Transformation."

245 **The American military wanted:** Author interview with Steve Archibald.

5.

246 **Luis Rueda dispatched the first:** Author interview with Luis Rueda.

246 **He slept in his office:** Author interview with Luis Rueda.

247 **Then, finally, a message:** Bob Woodward, *Plan of Attack* (Simon & Schuster, 2004), 140–42.

248 **Tom began handing out:** Author interview with Yousif Ismael.

248 **One came with a vial:** Woodward, *Plan of Attack,* 142–43.

248 **when Rueda got a cable:** Author interview with Luis Rueda.

249 **They'd pierce their faces:** Erik Peper, "Pain as a Contextual Experience," Townsend Letter, November 1, 2015.

249 **The kicker was that Saddam:** Quil Lawrence, *Invisible Nation: How the Kurds' Quest for Statehood Is Shaping Iraq and the Middle East* (Walker, 2008), 163.

249 **Some Kasnazani, especially ones:** Author interview with Luis Rueda; author interview with Yousif Ismael.

249 **Rueda tried to suppress:** Author interview with Luis Rueda.

250 **He watched Tommy Franks:** *Tommy Franks, American Soldier* (Regan Books, 2004), 266.

250 **Maybe troops felt:** Author interview with Luis Rueda.

251 **The Kasnazani tip looked:** Woodward, *Plan of Attack,* 300.

251 **Under pressure from the brothers:** Woodward, 211–12.

251 **The Kasnazani were everywhere:** Woodward, 212.

252 **Tom was doling out a million:** Tenet, *At the Center of the Storm,* 389–90.

252 **The team took to calling him:** Edward Wong, "As Sunni Divisions Widen, Iraq's Sufis Are Under Attack," *New York Times,* August 21, 2005.

252 **unprecedented, extraordinary intelligence bonanza:** Woodward, *Plan of Attack,* 143.

252 **He'd always known he had:** Author interview with Luis Rueda.

6.

254 **Kathy Kelly was trying:** Author interview with Kathy Kelly.

255 **She toured a music school:** Kathy Kelly, *Other Lands Have Dreams: From Baghdad to Pekin Prison* (CounterPunch, 2005), 9–11.

256 **And she did, in front:** Author interview with Kathy Kelly.

256 **She continued her work:** Ed Kinane, "Kathy Kelly: Glimpses of a 'Pious Pisshead,'" *Peace Council,* June 2005.

256 **Here in Baghdad, she could:** Author interview with Kathy Kelly.

256 **The family that ran her hotel:** Kelly, *Other Lands Have Dreams,* 62.

7.

258 **Best case, civilians:** Woodward, *Plan of Attack*, 329–31.
258 **Franks addressed the president:** "Iraq, Computer Modelling in Collateral Damage Estimates and Choice of Weapons," IHL in Action, undated.
258 **Even availing themselves:** Hemann, "Case Study Report: The Paveway Program Transformation."
258 **As Franks spoke, Bush:** Woodward, *Plan of Attack,* 331.
258 **Across the river at Langley:** Author interview with Luis Rueda.
259 **A "time of our choosing":** T. Michael Moseley, "Operation Iraqi Freedom—by the Numbers," Richard G. Trefry Archives, American Public University System, April 30, 2003, 15.
260 **"In the very near future":** Woodward, *Plan of Attack,* 374.
260 **So Rueda at least had that:** Author interview with Luis Rueda.
260 **President Bush made his way:** Franks, *American Soldier*, 8–10.
260 **Moseley spoke up from CENTCOM:** Woodward, *Plan of Attack,* 378.
261 **Bush turned to Donald Rumsfeld:** George W. Bush, *Decision Points* (Crown, 2010), 223.
261 **On that order, highly skilled:** Woodward, *Plan of Attack,* 379.
261 **Bush left the Situation Room:** Bush, *Decision Points*, 223.
261 **"I know I have taken":** Bush, 224.

8.

263 **Until now, he'd maintained:** Woodward, *Plan of Attack*, 300.
264 **He started asking around:** Tenet, *At the Center of the Storm*, 391–94.
264 **The CIA sent coordinates corresponding:** "Iraq Missile Attacks Missed Real Targets," *ABC News*, August 30, 2004; author interview with Luis Rueda.
265 **Almost immediately, Rokan:** Woodward, *Plan of Attack*, 372.
265 **The unassuming telecom officer:** Woodward, 374.

9.

266 **Rueda was snoring:** Author interview with Luis Rueda.
266 **"Come with us!":** Tenet, *At the Center of the Storm*, 594.
267 **One of the pictures showed:** Tenet, 593.
267 **At the Pentagon, Rumsfeld:** Tenet, 594.

10.

268 **Rueda crowded into the room:** Author interview with Luis Rueda.
268 **Before they could start, President:** Richard B. Myers, *Eyes on the Horizon: Serving on the Front Lines of National Security* (Threshold Editions, 2009), 235–36.
268 **Tenet got right to it:** Bush, *Decision Points*, 254.
269 **"This is really good":** Woodward, *Plan of Attack*, 383–84.
269 **He asked how good:** David Martin, "Ex-CIA Officer on the Strike That Could Have Averted Iraq War," *CBS News*, March 19, 2013.
269 **"We'll never get a hundred percent":** Woodward, *Plan of Attack*, 383–84.

269 **Bush nodded. He turned:** Walter L. Perry et al., *Operation Iraqi Freedom: Decisive War, Elusive Peace* (Rand Corporation, 2015), 59.
269 **Myers said cruise missiles:** Woodward, *Plan of Attack*, 384–87.
269 **This wasn't a target:** Bush, *Decision Points*, 254; Rumsfeld, *Known and Unknown*, 459.
270 **According to Rokan, a yellow taxi:** Myers, *Eyes on the Horizon*, 237–38.
270 **Plus, a Dora Farms staff member:** Author interview with Luis Rueda.
270 **When the asset followed:** Woodward, *Plan of Attack*, 386.
270 **Tom compiled the latest:** Tenet, *At the Center of the Storm*, 393–94.
270 **Rumsfeld, Cheney, Bush, Rice:** "Regime Change," "The Iraq War" (documentary series,) BBC, May 2013.
270 **Rueda dragged his fingers:** Woodward, *Plan of Attack*, 386–87.
271 **If indeed the intelligence Rueda's:** Adam J. Hebert, "The Baghdad Strikes," *Air & Space Forces*, July 2003.

11.

273 **The test range ground crew:** Paul F. Crickmore, *Lockheed F-117 Nighthawk Stealth Fighter* (Osprey Publishing, 2014), 37–38; author interview with Steve Archibald.
273 **The Raytheon team briefed:** Crickmore, *Lockheed F-117 Nighthawk*, 37–38.
273 **Both bombs began gliding:** Andreas Parsch, "Paveway III," Directory of U.S. Military Rockets and Missile, August 2008; "Stealth Fighters Use New Munitions to Hit Baghdad," Navy.com, March 22, 2003.

12.

275 **In a building near the flight line:** Crickmore, *Lockheed F-117 Nighthawk*, 39.
275 **The two pilots said prayers:** Emily A. Kenney, *Bombs over Baghdad*, 49th Wing Public Affairs, March 18, 2015.
276 **In Washington, Luis Rueda:** Author interview with Luis Rueda.
277 **The president drew everyone's attention:** Rumsfeld, *Known and Unknown*, 460; Myers, *Eyes on the Horizon*, 240; Martin, "Ex-CIA Officer on the Strike That Could Have Averted Iraq War."
277 **"We should take the strike":** Martin, "Ex-CIA Officer on the Strike That Could Have Averted Iraq War."
277 **He took Cheney:** Woodward, *Plan of Attack*, 391.
277 **He went behind the *Resolute* Desk:** Tenet, *At the Center of the Storm*, 393.
277 **"OK, let's go":** John J. Lumpkin, "Stealth Fighters Target Saddam on First Night of War," Associated Press, April 12, 2003.

13.

278 **In the middle of the night, she'd received:** Kelly, *Other Lands Have Dreams*, 60.
278 **Saddam was on TV:** "U.S. Bombs Iraq, Hunts for Saddam," *Honolulu Advertiser*, March 20, 2003.
278 **On the second night:** Rajiv Chandrasekaran, "In the Dark over Power Outage," *Washington Post*, April 16, 2003.
278 **In the hotel bomb shelter:** Ramzi Kysia, "And Then the Bombs Began to Fall," *Voices in the Wilderness*, April 4, 2003.
279 **By the third night the explosions:** Kelly, *Other Lands Have Dreams*, 60.

279 **When the pace of explosions:** Author interview with Kathy Kelly.
281 **The convoy rumbled:** Kathy Kelly, "The Bravest Woman I Know," *Progressive*, December 2013.
281 **Courage for Peace:** Kathy Kelly, "An Eyewitness to the Horrors of the US 'Forever Wars' Speaks Out," *National Catholic Reporter*, January 3, 2020; Kathy Kelly, "Reparations and Healing Needed After Iraq War," *Progressive*, March 9, 2013.
282 **As they ate and drank:** Kelly, "An Eyewitness to the Horrors."
284 **She went back to Hisham:** Kelly, *Other Lands Have Dreams*, 10.

14.

285 **Then there'd been more:** Myers, *Eyes on the Horizon*, 240.
285 **When Saddam appeared on television:** Press briefing by White House Press Secretary Ari Fleischer, March 20, 2003.
286 **Rueda soon learned:** Tenet, *At the Center of the Storm*, 394–95.
286 **In the last hours of Bush's:** Kevin M. Woods et al., "Iraq Perspectives Project: A View of Operation Iraqi Freedom from Saddam's Senior Leadership," Joint Center for Operational Awareness, 2006.
286 **So among the casualties:** Tenet, *At the Center of the Storm*, 394–95.

15.

289 **In the newspapers, American commanders:** Patrick E. Tyler, "U.S. Forces Take Control of Baghdad," *New York Times*, April 10, 2003.
290 **He writes a memo:** Albert R. Hunt, "Killing Terrorists, Creating More," *New York Times*, April 16, 2013.
290 ***The New England Journal of Medicine*:** Iraq Family Health Survey Study Group, "Violence-Related Mortality in Iraq from 2002 to 2006," *New England Journal of Medicine*, January 2008.
290 **So many are buried without:** Gilbert Burnham et al., "Mortality After the 2003 Invasion of Iraq: A Cross-Sectional Cluster Sample Survey," *Lancet*, October 2006.
290 **Kathy and her team stay:** Kelly, "Lessons Learned in the Bucca Camp," *Stringer*, September 20, 2014.
293 **Somewhere a flag snaps:** Kelly, *Other Lands Have Dreams*, 89.
293 **Anyway: the camp is closed:** Kelly, "Lessons Learned."
293 **Kathy lingers, considering:** Kelly, *Other Lands Have Dreams*, 89.
293 **When Kathy walks in:** Kelly, 90.
294 **Olive-skinned and apprehensive:** Ruth Sherlock, "How a Talented Footballer Became World's Most Wanted Man, Abu Bakr al-Baghdadi," *Telegraph*, November 11, 2014.
295 **Ibrahim is developing the sense:** "Timeline: The Life and Death of Abu Bakr al Baghdadi," Wilson Center, October 28, 2019.
295 **He knows enough to talk:** Terrence McCoy, "How the Islamic State Evolved in an American Prison," *Washington Post*, November 4, 2014.

BOOK VII: SALMAN

1.

299 **In the mid-2000s:** Author interview with Steve Roemerman.

300 **The Department of Defense had computers:** Vint Cerf, "A Brief History of the Internet & Related Networks," *Internet Society.*

300 **TI was involved:** Caleb Pirtle, *Engineering the World: Stories from the First 75 Years of Texas Instruments* (Southern Methodist University Press, 2005), 144, 172–73.

300 **The mandate from Steve's:** Pirtle, *Engineering the World,* 146.

300 **TI had highly classified contracts:** Author interview with Steve Roemerman.

300 **There were TI teams working:** Nick Gromicko, "The History of Infrared Thermography," International Association of Certified Home Inspectors (InterNACHI), accessed February 13, 2024; Steven Leibson, "A Brief History of the Single-Chip DSP, Part II: After Two Solid Decades of Growth, FPGAs Became the Apex Predators of the DSP Jungle," *EE Journal,* September 8, 2021; F. R. Bailey, "Synthetic Aperture Radar Functional Diagram," Technical Report 3B-05, 1975, 306–8.

300 **The early obstacles to GPS:** "GAMBIT 1 KH-7 Film Recovery Vehicle," National Museum of the United States Air Force.

301 **TI was one of the companies:** James Estrin, "Kodak's First Digital Moment," *New York Times,* August 12, 2015.

301 **Whenever he asked project:** Author interview with Steve Roemerman.

302 **It was 1993, and Bill Clinton's:** William J. Clinton Presidential History Project Interview with R. James Woolsey, January 13, 2010.

303 **Karem had a company:** Richard Whittle, "The Man Who Invented the Predator," *Smithsonian,* April 2013.

303 **Woolsey purchased five:** Chris Woods, "The Story of America's Very First Drone Strike," *Atlantic,* May 30, 2015.

303 **The Defense Department caught:** Walter J. Boyne, "How the Predator Grew Teeth," *Air & Space Forces,* July 1, 2009.

303 **Since the prototype's new mission:** Chris Woods, "The Story of America's Very First Drone Strike: The CIA's Then-Secret Weapon Missed Taliban Leader Mullah Omar, Starting a Bureaucratic Fight That Has Lasted 14 Years," *Wired,* December 15, 2015.

303 **an early satellite antenna:** Timothy M. Cullen, "The MQ-9 Reaper Remotely Piloted Aircraft: Humans and Machines in Action," United States Air Force, PhD dissertation, MIT, 2011, 206.

303 **When the first Predator flew:** Frank Strickland, "The Early Evolution of the Predator Drone," *Studies in Intelligence,* March 2013.

304 **"What are we looking at?":** William J. Clinton Presidential History Project Interview with R. James Woolsey, January 13, 2010.

305 **Drone operators could identify:** Boyne, "How the Predator Grew Teeth."

305 **The solution was obvious:** Arthur Holland Michel, "How Rogue Techies Armed the Predator, Almost Stopped 9/11, and Accidentally Invented Remote War," *Wired,* December 17, 2015.

305 **In August 1998, a truck packed:** "Report of the Accountability Review Boards: Bombings of the US Embassies in Nairobi, Kenya and Dar es Salaam, Tanzania on August 7, 1998 ", Federation of American Scientists, January 1999.

305 **When the attacks were traced:** Woods, "America's Very First Drone Strike."
305 **There was an airstrip bordering:** Richard Whittle, *Predator: The Secret Origins of the Drone Revolution* (Henry Holt, 2014).
306 **Predator was given an upgraded:** Michel, "How Rogue Techies Armed."
306 **Their first experiments tried equipping:** Richard Whittle, "Hellfire Meets Predator: How the Drone Was Cleared to Shoot," *Smithsonian*, March 2015.
306 **As the United States prepared:** Jeremy Scahill, "Germany Is the Tell-Tale Heart of America's Drone War," *Intercept*, April 17, 2015.
307 **Just hours before:** Roger Connor, "The Predator, a Drone That Transformed Military Combat," National Air and Space Museum, March 9, 2018.
307 **In the early days:** Woods, "America's Very First Drone Strike."
307 **Soon an updated version:** U.S. Air Force Fact Sheet: History of the 432d Wing and 432d Air Expeditionary Wing, September 2019.
308 **The United States began exporting drones:** Jean-Baptiste Jeangène Vilmer, "Proliferated Drones: A Perspective on France," Center for a New American Security, May 2016.

2.

309 **The bookish leader who'd accumulated:** Seth G. Jones et al., *Rolling Back the Islamic State* (RAND Corporation, 2017).
309 **When his fighters captured:** "Preventing the Reemergence of Violent Extremism in Post-IS Iraq and Syria," Center on International Cooperation, New York University, March 2021, 10.
309 **It was no longer an Islamic:** Graeme Wood, "What ISIS's Leader Really Wants," *New Republic*, September 1, 2014.

3.

311 **Cameron fancied himself:** David Cameron, "Time to Be Counted," *Guardian*, March 17, 2003.
311 **He'd made a point:** Nicholas Watt, "David Cameron Outlines Foreign Policy Philosophy—but Don't Call It a Doctrine," *Guardian*, September 21, 2011.
312 **The Lockerbie Bombing:** "Libyans Sentenced for French Bombing," *BBC News*, March 10, 1999.
312 **terror groups in South Asia:** Judy Mandel, "Libya Under Qadhafi: A Pattern of Aggression," Ronald Reagan Library.
312 **extravagant arms purchases from Russia:** Andrew Feinstein, "Where Is Gaddafi's Vast Arms Stockpile?," *Guardian*, October 26, 2011.
312 **began developing weapons of mass:** "Chronology of Libya's Disarmament and Relations with the United States," Arms Control Association, January 2018.
312 **And for all Prime Minister Cameron's:** Nicholas Watt, "David Cameron on Libya: We Are Not the 'Pull Up the Drawbridge Generation,'" *Guardian*, October 2, 2011.
312 **After years of the quagmire:** Ryan Lizza, "The Consequentialist: How the Arab Spring Remade Obama's Foreign Policy," *New Yorker*, April 25, 2011.
312 **To the surprise of many:** Watt, "David Cameron on Libya."
312 **He assembled a small coalition:** Patrick Wintour and Nicholas Watt, "David Cameron's Libyan War: Why the PM Felt Gaddafi Had to Be Stopped," *Guardian*, October 2, 2011.

313 **Almost immediately, British military:** Todd R. Phinney, "Reflections on Operation Unified Protector," *Joint Force Quarterly*, April 1, 2014.

4.

314 **Salman Abedi and Abdalraouf Abdallah:** "Friend of Manchester Arena Bomber Loses Parole Bid," *BBC*, September 30, 2024.
314 **Kicked out of the house:** Josh Halliday, "Mohammed Abdallah, Whose Brother Was Close Friends with Salman Abedi, Travelled to Syria in 2014," *Guardian*, December 7, 2017.
314 **At sixteen his proportions:** Esther Addley et al., "Salman Abedi: From Hot-Headed Party Lover to Suicide Bomber," *Guardian*, May 26, 2017.
314 **He tried shoplifting:** Josh Halliday, "Jihadist with Links to Manchester Bomber Is Guilty of Fighting for Isis," *Guardian*, December 7, 2017.
314 **The two were a few years:** Manchester Arena Inquiry, December 2, 2021.
315 **By the time Arab Spring:** Rory Smith and Ceylan Yeğinsu, "For Manchester, as for Its Libyans, a Test of Faith," *New York Times,* May 25, 2017.
315 **When Libyan families in Manchester:** Jamie Doward et al., "How Manchester Bomber Salman Abedi Was Radicalized by His Links to Libya," *Guardian*, May 28, 2017.
316 **Salman was a few months:** Manchester Arena Inquiry, November 25, 2021.
316 **Salman was assigned:** Daniel De Simone, "Manchester Arena Inquiry: Salman Abedi 'Trained with Islamist Militia,'" *BBC News*, December 8, 2020.
316 **He did his best:** Manchester Arena Inquiry, November 22, 2021.
316 **They tried to learn how:** "Libya: Examination of Intervention and Collapse and the UK's Future Policy Options," House of Commons, January 19, 2016.
317 **With less and less capacity:** Manchester Arena Inquiry, November 25, 2021.
317 **So much so that Gaddafi's son:** Jane Deith, "The Teenage Libyan Rebel from Manchester," *Channel 4*, June 11, 2011.
317 **Murals appeared on walls:** Katrin Bennhold et al., "'Forgive Me': Manchester Bomber's Tangled Path of Conflict and Rebellion," *New York Times*, May 27, 2017.
317 **There was pride to help:** Manchester Arena Inquiry, November 25, 2021.
317 **the British began deploying:** "RAF Air Strikes in Iraq and Syria: October 2016," Ministry of Defence, December 16, 2024.

5.

318 **War planners began talking:** "Targeting," U.S. Air Force, November 12, 2021.
319 **when the British Royal Air Force:** George Allison, "Paveway—The Bomb Used to Blast Houthi Targets in Yemen," *UK Defence Journal*, January 12, 2024.
319 **They developed a way:** "Paveway IV Dual-Guidance Bomb on Aircraft Integration Path," *Defense Industry Daily*.
320 **And to make sure the weapon:** "Paveway IV Dual-Guidance Bomb."
320 **To further minimize civilian casualties:** "Paveway IV," *Think Defence*, November 20, 2022.

6.

322 **In the skies above them:** "Gaddafi Targeted in Series of Precision Strikes," *Malta Independent*, May 12, 2011.

323 **Salman and Abdalraouf arrived:** "Gadhafi Loyalists Resist Rebels in Tripoli," NPR, August 22, 2011

323 **As NATO planes made:** Kareem Fahim and Mark Mazzetti, "Rebels' Assault on Tripoli Began with Careful Work Inside," *New York Times*, August 22, 2011.

323 **No one knew where he was:** "Libya Crisis: Col. Gaddafi Vows to Fight a 'Long War,'" *BBC News*, September 1, 2011.

7.

325 **He had scant intel:** "Death of a Dictator: Bloody Vengeance in Sirte," Human Rights Watch, October 16, 2012.

326 **Mutassim began gathering patients:** "Death of a Dictator."

326 **A little after 8 a.m.:** Tracey Shelton, "Gaddafi Dead: Rebels Recount Libyan Leader's Final Moments," *World*, August 1, 2016.

327 **One of the lead drivers:** Tim Gaynor and Taha Zargoun, "Gaddafi Caught Like 'Rat' in a Drain, Humiliated and Shot," Reuters, October 21, 2011.

329 **He managed a question:** Martin Chulov, "Gaddafi's Last Moments: 'I Saw the Hand Holding the Gun and I Saw It Fire,'" *Guardian*, October 20, 2012.

329 **United Kingdom had led a real:** Operation UNIFIED PROTECTOR Final Mission Stats, North Atlantic Treaty Organization, November 2, 2011.

329 **Heads of state celebrated NATO:** Ivo H. Daalder and James G. Stavridis, "NATO's Success in Libya," *New York Times*, October 30, 2011.

330 **He stood at a podium:** David Smith, "Cameron and Sarkozy Meet Libya's New Leaders in Tripoli," *Guardian*, September 15, 2011.

8.

331 **He looked for diversions:** Nazia Parveen, "Bomber's Father Fought Against Gaddafi Regime with 'Terrorist' Group," *Guardian,* May 24, 2017.

331 **He found a South Manchester:** Amy Walker, "Manchester Arena Bomber 'Disapproved of Brother's Drug Use,'" *Guardian*, February 11, 2020.

331 **His mother grew:** Manchester Arena Inquiry, November 22, 2021.

331 **He got dizzy and weak:** Manchester Arena Inquiry, December 8, 2020.

332 **He had the prime minister's speech:** Trevor Rayne, "NATO War on Libya—Eyes on the Prize," *Revolutionary Communist Group*, October 7, 2011.

332 **Abdalraouf had it in his head:** Manchester Arena Inquiry, November 25, 2021.

332 **When Abdalraouf was arrested:** Daniel De Simone, "Manchester Arena Inquiry: Prisoner in Touch with Bomber to Be Released," BBC, November 25, 2020.

332 **He'd probably run afoul:** "RAF Gunner and Paraplegic Man Jailed for Terrorism Offences," BBC, July 15, 2016.

332 **Salman found more trouble:** Duncan Gardham, "Manchester Arena Inquiry: Bomber Salman Abedi 'Addicted to Painkillers and Had Anger Management Classes,'" *Sky News*, December 8, 2020.

332 **Imams from local mosques:** Manchester Arena Inquiry, November 24, 2021.

9.

334 **The city of Sirte:** Hayes Brown, "ISIS-Linked Militants Reportedly Staged a Crucifixion in Libya," *BuzzFeed News*, August 18, 2015.

335 **When Salman returned to England:** John Scheerhout, "Hashem Abedi: All the Evidence in the Manchester Arena Trial, Day by Day," *Manchester Evening News*, August 20, 2020.

335 **He seemed less hotheaded:** Manchester Arena Inquiry, November 25, 2021.

335 **To other young Libyans:** Scheerhout, "Hashem Abedi."

336 **And at the mosque, worshippers:** Daniel De Simone, "Manchester Arena Inquiry: Bomber's Messages Held 'for Years' Before Attack," BBC, October 21, 2021.

336 **Salman used burner phones:** Rukmini Callimachi and Eric Schmitt, "Manchester Bomber Met with ISIS Unit in Libya, Officials Say," *New York Times*, June 3, 2017.

336 **And if Salman's commitment:** Bennhold et al., "'Forgive Me': Manchester Bomber's Tangled Path of Conflict and Rebellion."

336 **He learned that a friend:** Doward et al., "How Manchester Bomber Salman Abedi Was Radicalised by His Links to Libya."

336 **Then Abdalraouf's trial ended:** "Friend of Manchester Arena Bomber Loses Parole Bid," BBC, September 30, 2024.

336 **The day after Abdalraouf:** "Abdul Hafidah Murder: Gang Sentenced for Moss Side Killing," BBC, September 15, 2017.

337 **Salman felt he saw:** Stephen Hopkins and George Bowden, "Manchester Bomber Salman Abedi Had a 'Bomb-Making Workshop,' Reports Claims," *HuffPost*, May 26, 2017.

337 **Salman took more concrete steps:** "Phone Holding Clue to Manchester Arena Bomb Plot Was Seized at Airport, Murder Trial Told," *Sunderland Echo*, February 6, 2020; Scheerhout, "Hashem Abedi"; John Scheerhout, "An Empty Terraced House in Rusholme Was Used by Salman and Hashem Abedi for Deliveries of Bomb-Making Chemicals, Trial Hears," *Manchester Evening News*, February 17, 2020.

337 **He bought a hatchback:** "Manchester Attack: Bomber Abedi 'Built Device Alone at Flat,'" *BBC*, June 11, 2017.

337 **He had the chemicals shipped:** John Scheerhout, "Manchester Arena Bomber Salman Abedi and Brother Hashem Fled Scene of Car Crash in Fallowfield Weeks Before Attack, Court Told," *Manchester Evening News*, February 14, 2020.

10.

339 **Lieutenant Rick Hilton was living:** Author interview with Rick Hilton.

340 **She'll be gone soon too:** Author interview with Rick Hilton.

341 **Lockheed Martin had just reached:** "Lockheed 'Paragon' Challenges in PGMs," *Air & Space Forces*, March 7, 2017; "Lockheed Martin Delivers 100,000th Paveway II Plus Laser Guided Bomb," *Aerotech News*, August 31, 2017.

341 **Raytheon, for its part:** "Raytheon Awarded First International Contract for Paveway IV," RTX, April 11, 2014.

341 **Raytheon's biggest revenue quarter:** "Raytheon Revenue Boosted by Strong Missile System Sales," Reuters, April 28, 2016.

342 **issued a policy directive:** Thomas X. Hammes, "Autonomous Weapons Are the Moral Choice," Atlantic Council, November 2, 2023; David Hambling, "U.S. to Equip MQ-9 Reaper Drones with Artificial Intelligence," *Forbes*, December 11, 2020.

11.

343 **On the evening of May 22, 2017:** "Manchester Arena Inquiry, vol. 1: Security for the Arena Report of the Public Inquiry into the Attack on Manchester Arena on 22nd May 2017, Hon. Sir John Saunders, Chairman," June 2021.

12.

346 **Three days after one of the:** Harriet Williamson, "I'm Sickened by the RAF Bomb with 'Love from Manchester' Written on It—How Dare You Drop That in My City's Name," *Independent,* May 26, 2017.

346 **There was initial skepticism:** Sarah Ann Harris, "RAF Bomb Heading for Islamic State Target Pictured with 'Love from Manchester' Message Scrawled On," *HuffPost,* May 26, 2017.

Index

Abdallah, Abdalraouf, 314–17, 322–23, 332, 334, 336
Abedi, Salman, 314–17, 322–23, 331–38, 343–45
Abu Ghraib, 294–95
acquisition radar, 32
Adams, Ansel, 66
Aden, xi-xiii
Afghanistan
  bin Laden and, 305–6
  War of 2001–21, 248, 307, 318, 329
Africa, 113
Air Operation Command, 275
air-to-air missiles, 58
Albright, Madeleine, 207
Al-Fanar Hotel, 254, 278
Allied Force, Operation, 212, 219, 229, 318
Allies, 4, 9, 22, 24, 287
Al Qaeda, 245
Al Rashid Hotel, 179
altitude dials, 61
Al Uswis Air Base, 275
antiaircraft defenses, 23, 33–34, 44, 71, 72, 159
anti-radar bombs, 58
anti-submarine sonar program, 56
antiwar protests, 256
Anvil, Project, 22, 57
Apollo mission, 27
Apple, 118–19
Arabs, 90, 100, 102–3, 109–10, 114, 121–22. *See also specific countries*
Arab Spring, 309–10, 315, 331
Assad, Bashar Al-, 332
Atari, 119
Axis, 23
B-17 Flying Fortress, 45
B-25 bomber, 52
Baghdad, 162, 164–65, 168–83, 249, 254–57
  Amiriyah Public Shelter No. 25, 165, 172–76, 181–83, 269
  Gulf War of 1991 and, 169–73, 176–83
  Iraq War of 2003–11 and, 282–84, 289–92
Baghdad Equestrian Club, 174
Baghdadi, Abu Bakr Al- (Ibrahim Awwad Ibrahim Ali al-Badri), 294–96, 309–11, 334–35
Balkans, 199–203, 206, 217–18, 222, 303–5, 312
bang-bang steering, 157
Bay of Pigs invasion, 235–36, 253, 269
Belgrade, 220–21
Bell Labs, 25
Berlin, 99–104, 108–15, 117, 121–28, 132, 134–36
Berlin Wall, 110
  fall of, 135–36
Berrigan, Daniel, 151, 192
biological weapons, 197
Black Hole (Riyadh Special Studies Division), 161–62, 174–75
Boeing, 258
bomb, quest for steerable, 22–23, 45–46. *See also* lasers, bombs guided by; Paveway laser-guided bomb
bomb plane, Allies attempt to destroy Nazi V-3 bunker with, 3–4, 8–13, 22, 28
Bosnia, 192, 201
Braun, Wernher von, 3–4, 27–29
Britain. *See* United Kingdom

British intelligence, 136
British military, 318
British Parliament, 311
British Royal Air Force, 319–21
British Secret Service, 316
British Torpex, 3
Brussels, 335, 343
Bullpup missile, 46
"bunker buster" bomb, 159–60. *See also* Paveway III
Burnage Academy, 314–15
Bush, George H. W., 163, 199–200, 261–63
Bush, George W., 224, 228–32, 240, 242, 250–52, 258–63, 268–70, 276–77, 285–88, 312

calculators, handheld, 95, 118
Cameron, David, 311–12, 330, 332
Camp Buca, 292–96, 310
Cannon (TV show), 188–89
Casey, William, 98
Castro, Fidel, 236–37
Central Command (CENTCOM), 155, 161, 174, 260
Central Intelligence Agency (CIA), 96, 98, 104, 122–23, 125–26, 136, 174–75, 236–37, 302–6
  Counterterrorism Center (CTC), 232, 241
  Farm, 238, 241
  Iraq Operations Group, 228–33, 239–40, 246–52, 258–60, 264–68, 270, 285
  Latin America division, 227, 238–41
Chad, 114
Challenger space shuttle, 189–90
Chanaa, Ali "Alba," 103, 108–9, 114–15, 121, 123
CHECKMATE Division, 153–54, 161, 175
chemical weapons, 197
Cheney, Dick, 268, 270, 277
China, 31, 33, 92
Christian Peace Teams, 339
Clinton, William Jefferson "Bill," 134, 199–202, 206–7, 212, 219, 224, 228–29, 302–3
Cohen, Leonard, 280–81
Cold War, 96, 99–100, 191, 199, 204, 237
Communism, 100
Communist Party of East Germany, 112
computers, 118–19, 158, 188
consumer products, 118–19
Cranston, Alan, 134
Croatia, 201
Cronkite, Walter, 169
cruise missiles, 155, 269, 271
C-SPAN, 163
Cuba, 234–36

Dayton Accords, 202, 206
DB/ROCKSTAR, 252
D-Day, 9
Defense Department (DoD, Pentagon), 51, 74, 76, 153, 161, 204, 209, 245, 268, 300, 303, 341–42
Defense Intelligence Agency (DIA), 175
de Havilland DH.98 Mosquito, 11
Democratic Party, 199
Desert Storm, Operation, 155, 169, 184, 190, 199–200, 219, 228
Detachment 5, 51–57, 61, 64, 68, 73
digital imaging, 301–2
Disney, Walt, 27
Disneyland, 27
Dora Farms, 265–67, 269–70, 275–76, 285–87
Dornan, Robert, 134
Dragon's Jaw bridge, 18–20, 31, 43–47, 51–52, 55, 65, 69–72, 81–86, 92, 138–39, 159, 307, 340
drones, 302–8, 341–42. *See also* Predator
dual axis attacks, 46
Dulles, Allan, 246
dynamic targeting, 318–19

Eagle Claw, Operation, 97–98
East Berlin, 99–104, 108, 110–15, 117, 121, 124, 134–35
  Libyan embassy spies and, 90–91, 101–2, 111, 114, 121, 123
Eastern Bloc, 156, 238
East German Ministry of State Security "Stasi," 89–91, 99–104, 108–10, 112, 114–15, 121, 136
Eastman Kodak Company, 301
Eastwood, Clint, 165
Edwards Air Force Base, 273
Eglin Air Force Base, 51–53, 68, 73
8th Expeditionary Wing, 275
Einstein, Albert, 20

El Al Airlines, 106
El Dorado Canyon, Operation, 129, 154, 161, 184, 200, 219, 311, 315
elections of 1996, 201
electro-optical bombs, 46, 58
engagement radar array, 32–33
Eter, Musbah Abdulgasem, 90–91, 101, 103, 108–9, 112–15, 117, 123–24, 126, 134–37
Europe, 201, 343
ex-Nazi scientists, 22–23, 26–29, 57
"eye" of missile, 58–60, 63

F-111 Aardvark planes, 117, 130–31, 138–39
F-117A Nighthawk planes, 176–78
Facebook, 299
Fadi, 291–92
fiber-optic cable, 307
fins
  laser-guided bomb and, 58–59
  retractable, 118
firebombing of Japanese cities, 77
Florida, 4–5, 29, 221–22, 347–48
Ford, Kenneth, 132
48th Tactical Fighter Wing, 128–33
433rd Tactical Fighter Squadron "Satan's Angels," 43
France, 312, 330
  Nazi-occupied, 3–4, 11
Franks, Tommy, 250, 258, 260–62
French intelligence, 136

Gaddafi, Hana, 133
Gaddafi, Muammar Muhammad Minyar al-, 105–8, 113–14, 131–33, 311–17, 322–30, 332
Gaddafi, Mutassim, 325–26
Garrity, Major, 293
General Dynamics, 117
Germany. *See also* Berlin
  Nazi, 2–6, 8, 21–23, 45, 287
  post-WWII, 306–7, 336
Giannini probe, 61, 63, 73–74, 77
gliders, 62–63
global war on terror, 307
Goatee 01, 70, 81–86
GPS (global positioning system), 188–90, 243, 300, 318
Grande, Ariana, 344–45
Grumman Aerospace, 204
Guantánamo, 207
Gulf War of 1991, 166–83, 228

Hadley, Stephen, 270
Haggerty, Pat, 24–27, 188–89
Hanoi, 32, 38, 45
Harris, 282–83
Harvard University, 6
Havana, 234–35
Heat 34, 176–78
Hellfire missiles, 306, 308
Hillah cholera outbreak, 283
Hilton, Ann, 44, 47, 86, 340
Hilton, Richard "Rick," 17–18, 30–45, 47, 70–72, 81–86, 130, 138, 153, 159, 187, 339–41
Hiroshima, 290
Hisham, 255–56, 283–84
Hitler, Adolf, 3–4, 7–9, 11, 22, 28, 162, 170, 223
Hughes, Howard, 205
Human Rights Watch, xiii
Huntsville, Alabama, 22
Hussein, Saddam, 255, 278, 280, 283, 289, 295
  CIA desire to topple, post-Gulf War, 228–30, 232–34, 241, 248–52
  Gulf War of 1991 and, 170, 173–75, 181, 228–29
  invasion of Kuwait and, 150–51, 154, 161–65, 181
  Paveway III strike on, 258–71, 275, 285–86, 319

Imperial Japanese Army, 23
inertial navigation system (INS), 243
insulator, 25
integrated circuits, 119
Iranian hostage crisis, 98
Iraq, 161–62, 228, 339
  Gulf War of 1991 and, 150–52, 154–55, 164–67, 169–81, 192, 219
  ISIS and, 334
  War of 2003–11 and, 246–53, 259–60, 278–96, 307, 309–10, 312, 318, 329
Iraqi Fedayeen Saddam, 251
Iraqi Freedom, Operation, 261, 272, 287
Iraqi General Intelligence Department, 175
Iraqi Ministry of Culture, 171, 180

Iraqi Republican Guard, 230
Iraqi Special Security Forces, 251
Iraqi Special Security Organization
Telecommunication Station, 263
Islamic State (*formerly* ISIS), 296, 309–11, 334–37, 343
Islamists, 332
Israel, 109, 113
Munich Olympics team killings and, 96
Rome and Vienna airport attacks of 1985 vs., 106–7

Japan, WWII and, 23–24, 77, 287
Japanese Mitsubishi Zero, 24
JDAM guided bomb, 258
Jews, 100
Jihad, 291–92
Johnson, Dick, 56–57, 61–62, 65–66, 68–69, 75, 77
Johnson, Lyndon B., 79–80
Joint Chiefs of Staff (JCS), 98, 155, 268
Joint Standoff Weapon (JSOW), 210
Jordan, 180, 261, 291

Kappes, Steve, 232
Karem, Abe, 303
Karl, 147–48, 151
Karma 51, 139
Kasnazani network, 249, 251–53, 258–59, 263–65, 269–70, 275, 286–87
Kelly, Kathy
father's illness in Chicago and, 222–24
imprisonment of, for breaking into nuclear missile silo, 143–50, 153
Iraqi invasion of Kuwait and, 150–52
trip to Canada, 339
trip to Iraq of 1991 and, 163–64, 166–68, 170–72, 179–83
trip to Iraq of 2003–4 and, 254–57, 278–84, 289–96, 309
trip to Sarajevo and, 191–96
Kelly, (Kathy's father), 164, 222–23
Kelly, (Kathy's mother), 163–64, 170
Kennedy, Joe, Jr., 3–13, 28, 46, 57, 184, 306
Kennedy, Joe, Sr., 6–7
Kennedy, John F., 6–8, 28, 30, 236, 253
Kennedy, Rose, 7
Kenya, U.S. embassy bombing of 1998 and, 305
Khrushchev, Nikita, 236
Kiki. *See* Nikolic, Ljubica
King Khalid Air Base, 176
Kosovo, 206–7, 212, 219, 229, 318
Kosovo Liberation Army (KLA), 206–7
Kurds, 230, 233, 239–41, 247–48
Kuwait, Iraqi invasion of, 150–51, 154–55, 163–67, 181, 228

La Belle nightclub bombing, 109, 126–27, 130, 132–33, 135–37
Laden, Osama bin, 305–7
Lake, Tony, 201
Lakenheath Air Base, 129–31
laminar flow airfoil, 62
Lancet, 290
laser designators, 77–78, 305
laser frequencies, 119
lasers, 20–21
bombs guided by, 52, 55–69, 73–86 (*see also* Paveway laser-guided bomb)
drones and, 305
Lebanon, 312
le Carré, John, 99
Lee, Bruce, 165
LeMay, Curtis E., 76–78
Lexington Federal Prison, 143
Libya
aftermath of death of Gaddafi in, 333–36
Arab Spring and Paveway strikes on Gadaffi and, 309, 311–17, 322–33
assassination plot vs. Reagan and, 96
Berlin bombing of 1986 and, 90–91, 96–99, 101–14, 121–27, 135–37
British embassy in, 105–6
London attacks of 1984 and, 105
Munich Olympics attacks of 1972 and, 96
Paveway II strike of 1986 on, 120, 130–34, 138, 151, 153–55, 184, 219
Reagan and preemptive strike plan vs., 96–98, 105–8, 116, 120, 129–35, 138
Rome and Vienna airport attacks of 1985 and, 106–7
Sudan attack of 1973 and, 96
Vienna attack of 1975 and, 96
Libyan immigrants, 315, 336–37
Libyan People's Bureau, 101, 114, 134

Life, 27
Lockerbie, Scotland, bombing, 135, 311–12
Lockheed Martin, 159, 204, 209, 211, 242, 258, 341
London, 315
  Blitz and Nazi rocket attacks on, 3–4, 7, 9, 223
  Libyan embassy attacks and, 105
Lorraine, 149
Los Angeles Times, 134
Lučani, Serbia, xiii, 197–98, 212–18, 220–21, 348

magnetic anomaly detector, 23–24
Malta, German embassy in, 136
Manchester, England, 315–17, 331, 336–38
Manchester Arena bombing of 2017, 343–45
Manilow, Barry, 66
Martin Marietta, 20–21, 204
Marxists, 100
McLaughlin, John, 228, 232
Messerschmitt plane, 24
microchips, 119, 158, 300. *See also* silicon semiconductor chips
Mielke, Erichs, 102
MiG fighter jets, 30–33, 38, 43–44, 51, 70–71, 83, 85, 92
Mike, 32, 34–35, 37–40, 42
Miladhah (night manager's daughter), 257, 278–79, 281
Milošović, Slobodan, 200–202, 206–7, 212, 219–20, 312, 318, 329
Minuteman missile, 27
Mitchell, Joni, 233
mobile phones, 243
moon landings, 27–28
Moseley, Buzz, 260
Mostar, 193, 196–98, 212, 304
Muhammad, Prophet, 310
Munich Olympics killings, 96
Mustafa, 101–3, 110, 121
Myers, Richard, 268–69, 271

NASA, 188
NASCAR, 303
National Airborne Operations Center, 128
National Security Agency (NSA), 100
National Security Council (NSC), 97–98, 105–6, 128, 259–60
  Decision Directive 138, 105–7
Nayirah (Kuwaiti ambassador's daughter), 150–51, 163, 181
Nellis Air Force Base, 116–17, 129
New England Journal of Medicine, 290
New York Police Department, 166
New York Times, 329
New York Times Magazine, xi
Nikolic, Ljubica "Kiki," 198, 213–18, 220–22, 347–49
Nikolic, Radica, xiii, 198, 215–18, 220–22, 347–58
nitrogen mustard gas, 197
Nobel Peace Prize, 339
Nobel Prize in Physics, 119
North Africans, 114
North American Aviation, 52, 56, 60, 65, 76
North Atlantic Treaty Organization (NATO)
  Balkans and, 201–2, 208, 212, 217–18, 312
  drones and, 308
  Islamic State and, 310, 332, 334
  Libya and, 113, 312–13, 316–17, 323–24, 329–30
Northrup Grumman, 204
North Vietnam, 17, 30–32, 41, 43–44, 46, 70–71, 79–82, 92
nuclear missile silos, 143–49

Obama, Barack, 312
oil industry, 23–23
Oman, 309
OPEC, Vienna terrorist attack vs., 96

P-51 Mustang jet, 52
Palestinians, 114–15
Paperclip, Operation, 22
Paris, 335, 343
Patton, George S., 9
Paveway laser-guided bomb
  Iraq War and, 258, 289–90
  Kosovo and, 318
  Libya attacks and Gadaffi killing and, 313, 322–24, 327–29, 334
  Predator drones and, 305, 308, 324, 327–29, 334, 341–42

Paveway laser-guided bomb (*cont.*)
  Raytheon takeover of, 204–5, 208–11, 242–44
  sales of, 245, 341
  satellites and, 187–90
Paveway I, xiii, 158
  bang-bang steering and, 157
  development of, 52, 55–69, 73–86, 93, 95
  manufacturing of, 93–95, 116–17
  sales of, 116
  Vietnam War and Dragon's Jaw and, 81–86, 92–93, 138
Paveway II, 158, 209, 244
  bang-bang steering and, 157
  development of, for "pull-up," 117–20, 156
  Hilton's foreboding about, 138–40
  integrated circuit microchip and, 119
  Libya attacks and, 128–31, 132–34, 138–40, 155–56, 323
  sales of, 156
Paveway III, 187, 319, 322–23
  Balkans and, 202, 207–8, 219, 242–43
  development of, 156–60
  enhanced "bunker buster" with INS and GPS and, 243–45, 272–76, 285, 319
  Gulf War and, 169–70, 175–78, 184
  steering and, 157
Paveway IV, 346
  height of burst sensor and autonomy of, 320–21
  Libya attacks of 2011 and, 317, 319–22
PayPal, 299
Peace Camp, 339
Peace Teams, 166, 171, 180, 192, 223, 254
Pearl Harbor attacks, 23
Phantom fighter jet, 30, 40
Phantom of the Opera, The (musical), 280–81
photography, digitization of, 301
photon sensor lenses, 58–60, 78
pitot tube, 61
Polaroid, 66
Powell, Colin, 155, 254, 268
precision bombing, 31, 77, 95, 116, 139, 153–54, 158, 200–204, 207–8, 219, 229, 232–33, 242, 244, 250, 258, 264, 274, 276, 286–87, 289, 312, 318, 323, 330, 336. *See also* Paveway laser-guided bomb; *specific targets*
Predator drones, 303–8
  with Hellfire missiles, 306–8
  with Paveways, 308, 324
preemptive strikes, 98, 105, 116, 129–34
Project 1559, 29, 56, 64
Project Paveway, 78
proximity fuzes, 24
Puerto Rico, 7–8

Qatar, 275, 278

radar, 23, 32–33, 44–46, 51, 58
Radica. *See* Nikolic, Radica
Ramstein Air Base, 306–7
Raqqa, 309
Rather, Dan, 169
Raytheon Missile Systems, 205, 209–11, 242–45, 272–74, 317, 319–21, 341
Reagan, Ronald, 96–98, 105–7, 116–17, 128–29, 132–34
Rebelde (TV show), 221, 349
Red Crescent, 181–82
Red Cross, 294
Red Crown, 32–33
Redstone rocket program, 22, 27–29, 57
Rice, Condoleezza "Condi," 268–70
Ries, Cary, 356
Ries, Elkan (grandfather), xiii, 352–58
Ries, Shirley (grandmother), 352–58
rockets, 22, 26–28, 33
Roemerman, Steve, 299–302
Rokan, 265–67, 270, 286–87
Rome, airport attack of 1985 and, 106–7
Roosevelt, Elliott, 11–12
Rueda, Luis, 227–42, 246–53, 258–60, 262, 266–71, 275–77, 285–88
Rueda, (Luis's father), 235–37, 277
Rumsfeld, Donald, 250, 261–62, 267–68, 270, 290
Russia, post-Soviet, 228, 312. *See also* Soviet Union

Sarajevo, 191–96, 198–202, 206–7, 212
Sarasota, Florida, 347–48
sarin gas, 197
satellites, 187–90, 243, 300
Saturn V rocket, 27
Saudi Arabia, 161, 174, 176, 260, 275
Schwarzkopf, Norman, Jr., 155

Section 25 (Stasi unit), 102
semiconductors, 25–26. *See also* silicon semiconductor chips
September 11, 2001 attacks, 231–32, 245, 255, 293, 307, 311
Serbia, xiii, 198, 200–202, 206, 212, 218–21, 242, 318, 354
Serbian immigrants, 347
Serbian Military Technical Institute, 197
Shanksville, Pennsylvania, 245
Shultz, George, 97–98
Sickle, Jack, 56–59, 62, 77, 93
signal amplifier
  microchip, 119
  photon sensor, 58
  vacuum tubes, 22–23
silicon, 25–26
silicon semiconductor chips, 26–29, 52, 56, 58–59, 78–79, 93–95 116–19
Sirte, Libya, 324–27, 334
Skunk Works division, 159
smart bombs, 200
sonar, 23
South Vietnamese military, 80
Soviet Union (Russia), 30–31, 92, 156, 238. *See also* Russia, post-Soviet
  collapse of, 187, 191, 203, 224, 239
Space Race, 65
Springer Publishing Company, 109, 110
spy satellites, 300–301, 303
Srebrenica massacre, 201, 312
Stalin, Joseph, 162
Stalinists, 100, 103
Stardust nightclub, 109, 112
Stasi. *See* East German Ministry of State Security
stealth bombers, 159–60, 177, 187, 242–43, 275
Strategic Air Command, 39, 77
Stuchly, Wolfgang, 102–4, 108–12, 121
Sufi mystics, 248–49, 253, 259
surface-to-air missiles (SAMs), 33–36, 38, 40–41, 43–44, 46, 51, 56–58, 70–71, 83, 85, 92, 156, 169
Syria, 296, 309–10, 330, 332, 334

Tandy Radio Shack, 119
Tanzania, U.S. embassy bombing of 1998 and, 305
Tenet, George, 230–31, 267–68
terrain-following radar system, 117, 128, 139, 300
Texas Instruments (TI; *formerly* Geophysical Service Incorporated), 299–300
  bomb navigation satellites and GPS and, 187–90
  classified projects and, 300
  consumer products and, 95
  Defense Systems & Electronics Group, 138
  end of Cold War and, 204–5
  GPS and, 243–44
  imaging projects and, 301
  manufacturing and, 93–95, 116
  Paveway consultant at Nellis and, 116–17
  Paveway I smart bomb developed by, 52–65, 76, 78, 92–95
  Paveway II and, 128–29, 155–56
  Paveway III and, 157–60, 169
  Raytheon acquires Paveway from, 205
  semiconductor chips and, 19, 23–29
  terrain-following radar system and, 117
  Word retires from, 183–84, 203
Thailand, 41
37th Fighter Wing, 176
379th Air Expeditionary Wing, 275–76
Thuraya satellite phones, 248, 252, 266
Time, 27
Today show, 207
Tom, 241, 247–48, 250–52, 259, 264–67, 269–70
Tomorrowland, 27
transistor, 25
Treasury Department, 166
Tripoli, 131–32, 322–24, 331, 334
Tunisia, 309, 316
Turkey, 247–48
Tucson International Airport, 273
Tucson weapons factory, xiii
Twitter, 299

U-boats, 8, 24
Ubon Royal Thai Air Force Base, 43, 45, 86
Umm Qasr, 292
Unified Protector, Operation, 313, 329, 330

United Kingdom (UK), 7, 130, 135, 208, 311–15, 329–30, 336, 343
United Nations (UN), 167, 200, 254–56, 313
U.S. Air Force, 29, 46, 52–53, 56–57, 79, 92, 97, 116–17, 138, 157, 159, 188, 189, 273, 306, 341
    Aeronautical Systems Division, 51, 78
    Directorate of Warfighting Concepts, 154
    Limited War Department, 51
    Logistics Command, 76–77
    Systems Command, 51
U.S. Army, 20–21, 80, 97, 125, 261
U.S. Army Hospital, West Berlin, 109
U.S. Army-Navy joint operations, 46
U.S. Congress, 97, 129, 134, 150–51, 181
U.S. House of Representatives, 232
U.S. intelligence, 4
U.S. Marines, 155, 282–84, 291
U.S. Navy, xiii, 24, 32–33, 97, 155, 261, 285
U.S. Navy SEALs, 241
U.S. Senate, 163, 232

vacuum tubes, 22–23, 26–27
Vega 31 stealth bomber, 242–43
vengeance weapons, 3–4, 8–9, 22, 27, 45
    V-1 rocket, 3, 9
    V-2 rocket, 3, 9, 22
    V-3 rocket "supergun," 3–6, 8–9, 22, 45
Verena, 115, 123–24, 126–27
Vessey, John W., Jr., 98
Vienna
    International Airport, attacks of 1985, 106–7
    OPEC attacks of 1975, 96
Viet Cong, 43
Vietnam War, 17–19, 29–46, 52, 79–86, 92–95, 97, 130, 153–54, 256, 340
VX nerve gas, 197

Warden, John, 153–55, 161–62, 174
Washington Post, 229
Weaver, Tom, 21–24, 26–29, 57
Weinberger, Caspar "Cap," 97, 106–7, 128
West Berlin, 99–100, 104, 108–12, 114, 121–28, 132
West Germany, 136
White Goat (film), 213–14
White House Situation Room, 105–7, 128–29, 132, 258–60, 270
Whitman Air Force Base, 144–47
Wideman, Bill, 70–71, 81–86, 139
Wiegand, Rainer, 89–91, 99–104, 108–12, 121–22, 136–37
Woolsey, James, 303–6
Word, Weldon, 202
    Desert Storm images and, 184
    family of, and death of first wife, 54–55
    foresees digital imaging technology and autonomous strike ability, 301–2, 307, 341
    moves to Tyler with second wife, 183
    Paveway I laser-guided bomb developed by, with TI team, 19–25, 27, 52–68, 73–80, 93
    Paveway II developed by, for low altitude release, 117–20, 138
    Paveway III developed by, as "bunker buster" for stealth bomber, 155–60, 169, 187
    Paveway III enhancement with GPS and, 187–89, 243, 319
    Paveway manufacture, simplicity and low-cost mantra of, 93–95
    personality and appearance of, 54–55
    Raytheon takeover of Paveway and, 210–11
    retirement of, 203–5, 299, 341
    as TI vice president, 155
word processors, 95
World Trade Center, 245, 255
World War I, 223
World War II, 3–8, 21, 23–24, 45, 52, 77, 145, 163, 188, 204, 223, 315
World Wide Web, 300
Wright-Patterson Air Force Base, 64, 76–79, 202
wristwatches, 118

Yemen, xi-xiii, 309
YouTube, 299
Yugoslavia, 191–96. *See also* Balkans

About the Author

Jeffrey E. Stern is an award-winning journalist and the author of four books, including *The 15:17 to Paris*, which was adapted as a major motion picture by Clint Eastwood and Warner Bros., and *The Last Thousand: One School's Promise in a Nation at War*, an honorable mention for Best Book of the Year by *Library Journal.* He has been named a graduate fellow at the Stanford Center for International Conflict and Negotiation and a grantee of the Pulitzer Center Fellowship for Crisis Reporting. Stern's reporting has appeared in magazines such as *The New York Times*, *Vanity Fair*, and *The Atlantic.*